Christoph von Beeren

Social integration of macro-parasites in ant societies

Christoph von Beeren

Social integration of macro-parasites in ant societies

Proximate and ultimate mechanisms

Südwestdeutscher Verlag für Hochschulschriften

Impressum/Imprint (nur für Deutschland/only for Germany)
Bibliografische Information der Deutschen Nationalbibliothek: Die Deutsche Nationalbibliothek verzeichnet diese Publikation in der Deutschen Nationalbibliografie; detaillierte bibliografische Daten sind im Internet über http://dnb.d-nb.de abrufbar.
Alle in diesem Buch genannten Marken und Produktnamen unterliegen warenzeichen-, marken- oder patentrechtlichem Schutz bzw. sind Warenzeichen oder eingetragene Warenzeichen der jeweiligen Inhaber. Die Wiedergabe von Marken, Produktnamen, Gebrauchsnamen, Handelsnamen, Warenbezeichnungen u.s.w. in diesem Werk berechtigt auch ohne besondere Kennzeichnung nicht zu der Annahme, dass solche Namen im Sinne der Warenzeichen- und Markenschutzgesetzgebung als frei zu betrachten wären und daher von jedermann benutzt werden dürften.

Coverbild: www.ingimage.com

Verlag: Südwestdeutscher Verlag für Hochschulschriften GmbH & Co. KG
Heinrich-Böcking-Str. 6-8, 66121 Saarbrücken, Deutschland
Telefon +49 681 37 20 271-1, Telefax +49 681 37 20 271-0
Email: info@svh-verlag.de

Approved by: Munich, Ludwig-Maximilians University, Diss., 2012

Herstellung in Deutschland (siehe letzte Seite)
ISBN: 978-3-8381-3225-9

Imprint (only for USA, GB)
Bibliographic information published by the Deutsche Nationalbibliothek: The Deutsche Nationalbibliothek lists this publication in the Deutsche Nationalbibliografie; detailed bibliographic data are available in the Internet at http://dnb.d-nb.de.
Any brand names and product names mentioned in this book are subject to trademark, brand or patent protection and are trademarks or registered trademarks of their respective holders. The use of brand names, product names, common names, trade names, product descriptions etc. even without a particular marking in this works is in no way to be construed to mean that such names may be regarded as unrestricted in respect of trademark and brand protection legislation and could thus be used by anyone.

Cover image: www.ingimage.com

Publisher: Südwestdeutscher Verlag für Hochschulschriften GmbH & Co. KG
Heinrich-Böcking-Str. 6-8, 66121 Saarbrücken, Germany
Phone +49 681 37 20 271-1, Fax +49 681 37 20 271-0
Email: info@svh-verlag.de

Printed in the U.S.A.
Printed in the U.K. by (see last page)
ISBN: 978-3-8381-3225-9

Copyright © 2012 by the author and Südwestdeutscher Verlag für Hochschulschriften GmbH & Co. KG and licensors
All rights reserved. Saarbrücken 2012

„Die gelungenste Anpassungstactik ist aber jedenfalls die, dem übermächtigen Gegner als Freund sich anzuschließen und den Grundsatz zu befolgen: ‚Mit den Wölfen muss man heulen'. Wem das gelingt, dem ist eben durch die Gesellschaft seiner furchtbarsten Feinde ein mächtiger Schutz und eine reichgedeckte Tafel gesichert."

Erich Wasmann (1895)

Table of contents

Summary 1
Zusammenfassung 3
General introduction 5
    Ultimate mechanisms: Why are some myrmecophiles integrated and others are not? 9
    Proximate mechanisms: Why are some myrmecophiles integrated and others are not? 11
Chapter 1: Differential host defense against multiple parasites in ants 17
Chapter 2: Acquisition of chemical recognition cues facilitates integration into ant societies 41
Chapter 3: The social integration of a myrmecophilous spider does not depend exclusively on chemical mimicry 67
Chapter 4: On the use of adaptive resemblance terminology in chemical ecology 85
Summarized results 95
General discussion 97
    Ultimate mechanisms: Why are some myrmecophiles integrated and others are not? 98
        Ultimate mechanisms: Conclusion 100
        Ultimate mechanisms: Future directions 101
    Proximate mechanisms: Why are some myrmecophiles integrated and others are not? 102
        Origin of mimetic compounds 102
        The role of accuracy in chemical mimicry 106
        Proximate mechanisms: Conclusion 108
        Proximate mechanisms: Future directions 108
General outlook 109
References 111
Acknowledgements 117

Summary

Ant colonies are commonly parasitized simultaneously by several species. While some parasites are recognized and attacked by their ant hosts, others have apparently cracked the ants' recognition code and interact mainly peacefully with their hosts. Although such apparent differences in social integration among ant parasites have been described, the underlying mechanisms resulting in differential integration remain mostly unknown. Using *Leptogenys* army ants and their parasites, I studied ultimate mechanisms that might be responsible for differing integration levels by comparing the strength of host defence with the negative impact of parasites. In addition, I investigated proximate mechanisms of differing integration levels by evaluating the role of chemical deception by mimicry.

The interactions of several parasitic beetle species with their *Leptogenys* hosts revealed that particular species fed on host larvae, while others did not. The hosts' aggressiveness was enhanced towards brood-killing species, while non-predatory species received almost no aggression, resulting in social integration. Accordingly, the fitness costs of parasites likely influence the evolution of host defences against them in a multi-parasite situation.

The role of chemical mimicry has been investigated in detail for two kleptoparasites, namely the silverfish *Malayatelura ponerophila* and the spider *Gamasomorpha maschwitzi*. By analyzing the transfer of a chemical label from the host ants to the parasites, I empirically demonstrated for the first time that ant parasites are able to acquire mimetic compounds from their host. Additional biosynthesis of mimetic compounds seems unlikely in both parasites, since the concentration of each cuticular hydrocarbon decreased in individuals that were isolated from the host. In addition, a high accuracy in chemical host resemblance was shown to be beneficial for the social integration of both parasites. Reduced accuracy in chemical host resemblance resulted either in aggressive host responses towards the silverfish or elevated host inspection behaviour towards the spider. The degree of dependency on chemical mimicry to achieve social integration differed considerably between the two parasites, however.

Accordingly, the parasites' level of social integration is affected by ultimate mechanisms such as the negative impact on the host as well as by proximate mechanisms such as the degree of accuracy in chemical host resemblance.

Zusammenfassung

Ameisenkolonien werden häufig von verschiedenen Arten gleichzeitig parasitiert. Während manche Parasiten erkannt und attackiert werden, haben andere offensichtlich das Erkennungssystem der Wirtsameisen überlistet und interagieren zumeist friedlich mit den Wirten. Obwohl solch ausgeprägte Unterschiede in der sozialen Integration häufig beschrieben wurden, blieben die zugrundeliegenden Ursachen zumeist unbekannt. In meiner Dissertation untersuchte ich ultimate Gründe, welche für die Unterschiede in der sozialen Integration verantwortlich sein könnten. Hierzu verglich ich die Stärke der Wirtsabwehr mit dem negativen Einfluss der Parasiten auf ihre Wirte, südostasiatische Treiberameisen der Gattung *Leptogenys*. Außerdem untersuchte ich proximate Mechanismen der sozialen Integration, indem ich die Rolle chemischer Täuschung durch Mimikry beleuchtete.

Die Interaktionen zwischen verschiedenen parasitischen Käferarten und ihren *Leptogenys* Wirten zeigte, dass manche Käferarten die Brut der Wirte fraßen, während andere das nicht taten. Die Aggressivität der Wirte war gegenüber den Bruträubern erhöht, während Arten die keine Brut fraßen nicht attackiert wurden, so dass letztere ein hohes Maß an sozialer Integration erreichten. Folglich beeinflussen in einem Multi-Parasiten System die Fitnesskosten eines Parasiten wahrscheinlich das Ausmaß der gegen ihn gerichteten Wirtsabwehr.

Die Rolle der chemischen Mimikry wurde für zwei Kleptoparasiten untersucht, eine Silberfisch- und eine Spinnenart. Durch die Übertragung eines künstlichen Kohlenwasserstoffes von den Wirtsameisen auf die Parasiten konnte zum ersten Mal empirisch gezeigt werden, dass Ameisenparasiten in der Lage sind mimetische Substanzen von ihren Wirten zu erwerben. Beide Parasitenarten verloren mimetische Kohlenwasserstoffe, wenn sie von ihren Wirten getrennt wurden. Dies deutet darauf hin, dass sie selbst keine mimetischen Stoffe herstellen. Außerdem wurde gezeigt, dass eine hohe Genauigkeit der chemischen Ähnlichkeit zum Wirt für beide Parasitenarten vorteilhaft ist. Reduzierte Genauigkeit der chemischen Mimikry resultierte in aggressiver Reaktion der Wirte gegenüber den Silberfischen sowie in erhöhtem Inspektionsverhalten gegenüber den Spinnen. Die Abhängigkeit von chemischer Mimikry zur Erreichung sozialer Integration unterschied sich allerdings deutlich zwischen den beiden Parasiten.

Die Interaktionen zwischen Ameisenparasiten und ihren Wirten werden folglich sowohl von ultimaten Faktoren wie den Auswirkungen der Parasiten auf die Fitness der Wirte als auch von proximaten Faktoren wie der Genauigkeit der chemischen Ähnlichkeit der Parasiten zu den Wirten beeinflusst.

General introduction

"As well as being the causative organisms of major human and animal diseases, parasites often serve as elegant models for the study of fundamental biological phenomena."
J.D. Smyth (1994)

The spider Gamasomorpha maschwitzi *is one of the various species parasitizing the army ant* Leptogenys distinguenda.
© Christoph von Beeren

Symbioses, i.e. the permanent association between two or more distinct organisms during at least part of their life cycle (Goff 1982; Hughes et al. 2008), are common in all ecosystems on earth (Dimijian 2000). Symbioses are best considered as a continuum and range from mutualistic (both partners benefit) to parasitic associations (one partner benefits and the other is harmed). In the great majority of symbioses, one species parasitizes another species, e.g. by using it as a food source (Combes 2005). Indeed, parasitism is one of the most successful life strategies among eukaryotes, as measured by how often it evolved and how many parasitic species exist (Poulin and Morand 2000; de Meeus and Renaud 2002). Host-parasite interactions are often regarded as coevolutionary "arms races" in which the opponents exert reciprocal selection pressures on one another over long periods of time, often resulting in a dynamic equilibrium of fitness gains and losses (Dawkins and Krebs 1979; Thompson 2005). Since host-parasite interactions are of major importance as drivers of evolutionary processes (Thrall et al. 2007), they are ideal systems for the study of coevolution.

The major groups of social insects, i.e. ants, wasps, termites and bees, are known to harbour a great diversity of parasites, including nematodes, helminths, chelicerates, molluscs, collembolans, crustaceans and insects (Hölldobler and Wilson 1990; Schmid-Hempel 1998; Witte et al. 2002; Boomsma et al. 2005; Thomas et al. 2005). Ant-associated species are called 'myrmecophiles', meaning 'ant lover', from the Greek 'myrmex' (ant) and 'philos' (loving). The general definition of a 'myrmecophile' by Wilson (1971), i.e. any organism that depends on ants at least during part of its life-cycle, is rather broad and includes plants and animals as well as fungi and bacteria (Kronauer 2009). Since the impact of myrmecophiles on their host is often unknown, this definition puts emphasis on the persistence and specificity rather than on the quality of the association. However, a significant proportion of myrmecophiles appear to be parasitic (Howard et al. 1990a). In this thesis, I will use the term 'myrmecophile' to describe macroparasitic organisms only. Several factors may be responsible for the high species diversity of myrmecophiles. One factor may be the ants' ecological dominance in many terrestrial habitats in terms of abundance, biomass and energy turnover (Wilson 1990; Ward 2006). They accumulate considerable amounts of resources that can be of potential use for other organisms. Furthermore, ant colonies are expected to offer high microhabitat heterogeneity (Hölldobler and Wilson, 1990), which offers myrmecophiles the possibility to avoid interspecific competition by niche differentiation, thus sustaining a high species diversity within a single colony. Additionally, most ant species are characterized by long-living colonies, showing rather low extrinsic mortality rates which increases the probability that a given myrmecophile species will eventually colonize a given colony (Gotwald 1995; Hughes et al. 2008). Army ant colonies are the most extreme example of this, as new colonies are created by colony fission making them potentially immortal (Gotwald 1995). Interestingly, the greatest diversity of myrmecophiles, measured either per host species or per colony, is indeed found within the large societies of tropical army ants (Hölldobler and Wilson 1990; Gotwald 1995; Witte et al. 2008; Fig. I). Rettenmeyer et al. (2011) recently listed the enormous diversity of at least 300 species associated with the army ant *Eciton burchellii*.

(1) *Allopeas myrmecophilos*   (2) *Togpelenys gigantea*   (3) *Malayatelura ponerophila*   (4) *Gamasomorpha maschwitzi*

Figure I. A great diversity of myrmecophiles is found with army ant societies. This collage shows an emigration of the army ant *Leptogenys distinguenda* and some of its myrmecophiles. From left to right: (1) A myrmecophilous snail carried by an ant worker, (2) a staphylinid beetle follows the ants' pheromone trail, (3) a silverfish rides on ant pupae carried by an ant worker, and (4) a spider keeps contact to ants while joining the emigration. © Volker Witte

Under coevolution, one would expect that myrmecophiles adapt towards an efficient transmission between host colonies and successful exploitation, whereas host ants in turn adapt towards an avoidance of encounters or successful defence against parasitic myrmecophiles (Combes 2005; Cremer et al. 2007). While some myrmecophiles are in fact frequently attacked by ant workers, in a large number of cases myrmecophiles are integrated seamlessly, as if they were members of the society (Lubbock 1891; Wasmann 1895; Gösswald 1955; Kistner 1979; Hölldobler and Wilson 1990; Gotwald 1995). Several classifications have been suggested to depict the various myrmecophile-ant interactions (reviewed in Hölldobler and Wilson 1990). Here, I adopt the definition of Kistner (1979) describing myrmecophiles either as "integrated species" or "non-integrated species". While integrated species are incorporated into the host societies, eliciting a peaceful behaviour of their host towards them, non-integrated species attain no integration into the host society, eliciting aggressive host defence behaviour. Integrated species are generally found inside the ant nests staying in close contact to the host (Seevers 1965; Akre and Rettenmeyer 1966; Hölldobler and Wilson 1990). During emigrations of host ants, they typically move among the ant workers. Encounters between integrated myrmecophiles and host ants are frequent and mainly peaceful, in that myrmecophiles are often fed by ants, rub against host workers or larvae and are even sometimes groomed by their host ants. In contrast, non-integrated species are often found outside on the periphery of the ant nest, for example in refuse deposits or along ant trails (Akre and Rettenmeyer 1966; Hölldobler and Wilson 1990; Gotwald 1995). They typically follow host emigrations at the end, so that encounters between myrmecophiles and hosts are infrequent. Ants generally recognize and attack these myrmecophiles. As a consequence, non-integrated myrmecophiles often escape through quick movements, are morphologically protected and/or use other defence mechanisms.

Although different levels of integration among myrmecophiles were frequently described, their underlying mechanisms remain unknown in the majority of cases. In consequence, the question arose as to why some myrmecophiles are treated amicably while others are heavily attacked by their hosts. I used two different research approaches to elucidate the underlying mechanisms of differing integration levels. First, I studied the ultimate mechanisms probably dictating the integration levels by comparing the myrmecophiles' impact to the strength of host defence they receive (Chapter 1). Second, I investigated the proximate mechanisms of differing integration levels by observing the role of chemical deception by mimicry (Chapter 2 and 3).

Ultimate mechanisms: Why are some myrmecophiles integrated and others are not?

The recognition of non-self and the subsequent triggering of highly elaborate defence mechanisms are vital processes for most living beings (Combes 2005). In nature, hosts often have to defend themselves against several parasitic species simultaneously (Martens and Schon 2000; Rutrecht and Brown 2008; Rigaud et al. 2010). However, very few studies have investigated such multi-parasite situations thus far (Rigaud et al. 2010). The vast majority of studies on antagonistic associations of host-parasite or predator-prey systems have focused on one-to-one interactions (Laforsch and Tollrian 2004; Combes 2005). This approach, however, ignores the broader ecological context of multi-species associations, because evolutionary dynamics of one-to-one interactions strongly depend on the presence of other parasite/predator and host/prey species (Thompson 2005; Wolinska and King 2009; Rigaud et al. 2010). For multiply-parasitized hosts, the question arises as to whether the strength of host defence depends on the parasites' impact (Moore 2002). To the best of my knowledge, this question has not yet been addressed in a multi-parasite situation and was therefore one subject of my studies. If such a dependency exists, it could explain the different levels of integration found among myrmecophiles.

I studied the interactions of one particular group of myrmecophiles, staphylinid beetles (Coleoptera: Staphylinidae), with their army ant hosts so as to reduce the influence of taxonomic constraints. Studies on Neotropical staphylinid beetles of *Eciton* army ants revealed that both integrated and non-integrated species occur within this beetle family (Seevers 1965; Akre and Rettenmeyer 1966). Through convergent evolution, similar associations exist in Southeast Asia between staphylinid beetles and *Leptogenys* host ants (Kistner et al. 2003; Maruyama et al. 2010a; Maruyama et al. 2010b). I focused on five staphylinid beetle species (see Tab. I). Each beetle species only parasitizes one of two related army ant hosts, *Leptogenys distinguenda* or *L. borneensis*. The level of integration of each beetle species was assessed by studying the usual location of beetles in the ant colony, the beetles' behaviour during host emigrations and their interactions with host workers. The aggressiveness of *Leptogenys* ants is easy to evaluate and it is possible to determine the impact of staphylinid beetles on their host via feeding experiments (Witte et al. 2008; Witte et al. 2009). Accordingly, the potential fitness costs of beetles on their host were evaluated by their predation behaviour on host brood in isolation experiments (Chapter 1). Furthermore, the host defence was assessed by the ants' aggressiveness towards beetle individuals. I expected that the host defence, i.e. the aggressiveness of ant workers, would be stronger towards beetles that prey on the host, and less strong towards beetles that do not prey on the host. Accordingly, less costly (non-predatory) species are expected to achieve higher levels of social integration (Fig. II).

Figure II. Simplified scheme of a host that is parasitized simultaneously by two parasite species (P_1 and P_2). In coevolutionary arms races between multiple parasites and one host species, I expected the host defence to be elevated against more virulent parasites. Virulence is considered as the loss of host fitness due to parasites, which ranges from outright death to reduced fecundity. Parasites preying on host brood are expected to be more virulent than kleptoparasites (see discussion).

Proximate mechanisms: Why are some myrmecophiles integrated and others are not?

The pioneers of myrmecophile research noted that several species have somehow cracked the ants' recognition code, resulting in high integration levels (Lubbock 1891; Wasmann 1895). Several strategies allowing myrmecophiles to cope with their ant hosts have been described to date, such as protective morphological structures, behavioural adaptations, defensive or attractive glandular secretions, chemical or acoustical mimicry, and the complete lack of chemical recognition cues (Hölldobler and Wilson 1990; Gotwald 1995; Lenoir et al. 2001; Barbero et al. 2009; Stöffler et al. 2011). The first step for any myrmecophile individual is to find and successfully invade a host colony, while ants are expected to effectively recognize and defend themselves against intruders (according to Combes 2005). Since ants discriminate between colony members and alien species mainly on the basis of a particular group of chemicals, cuticular hydrocarbons (CHCs) (Blomquist and Bagnères 2010), many myrmecophiles evolved elaborate chemical strategies to deal with the ants' aggressive worker force (Akino 2008). The following chemical strategies may allow myrmecophiles to cope with their host: chemical mimicry (the mimic pretends to be an interesting entity), chemical crypsis (the mimic avoids detection through background matching), chemical masquerade (the mimic pretends to be an uninteresting entity), chemical hiding (suppression of any chemical recognition cues) or the use of ant deterrent/attractant chemicals (Lenoir et al. 2001; Akino 2008; Ruxton 2009; terms are used according to chapter 4). Since the terms describing chemical strategies are currently used inconsistently in chemical ecology literature, we presented a terminology that is consistent in itself and consistent with the use of terms in general biology (chapter 4). Among myrmecophiles, chemical mimicry by resembling host CHCs is probably the most frequent chemical strategy (Lenoir et al. 2001; Akino 2008).

The role of chemical mimicry as an integration mechanism was studied in two kleptoparasites, i.e. the silverfish *Malayatelura ponerophila* and the spider *Gamasomorpha maschwitzi*. Both species mimicked the CHCs of their *L. distinguenda* host workers and achieved high levels of social integration (Witte et al. 2009; Fig. III). They were found within ant nests, in which generally peaceful interactions with host workers occurred.

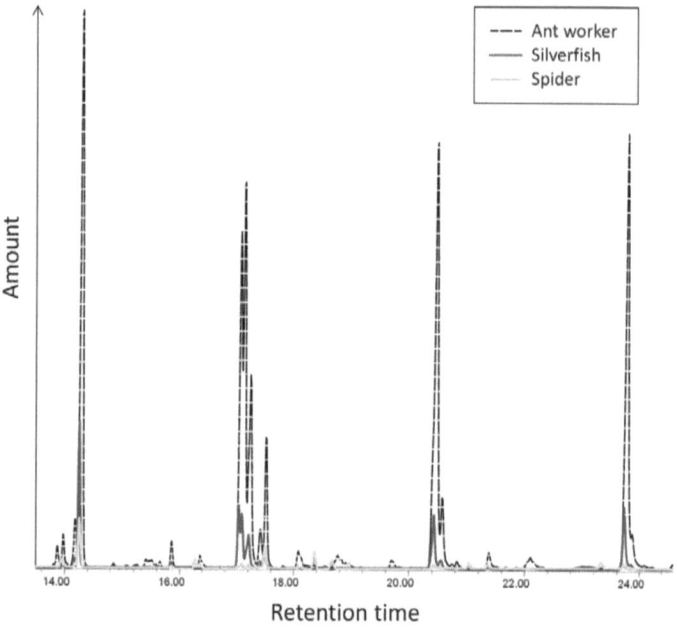

Figure III. Characteristic ion chromatograms from chemical profiles of a *L. distinguenda* host worker and two of its myrmecophiles, the silverfish *M. ponerophila* and the spider *G. maschwitzi*. Both myrmecophiles apparently mimic their hosts' cuticular hydrocarbons but to different degrees. For detailed information see Witte et al. (2009).

Two aspects of chemical mimicry were studied: the origin of mimetic compounds and the potential benefits for myrmceophiles on account of chemical mimicry. While some myrmecophiles probably acquire mimetic compounds through physical contact with the host, others are expected to biosynthesize them (reviewed in Akino 2008). In the majority of cases, however, the origin of mimetic compounds remains unclear, although a distinction between acquisition and biosynthesis of mimetic cues is useful as evolutionary consequences differ.

Mimetic and model cues are of identical origin if myrmecophiles acquire their compounds from the host ("acquired chemical mimicry" sensu chapter 4). In this case, coevolutionary arms races select for myrmecophiles with effective ways of acquiring host cues, e.g. through specific behaviours such as intense rubbing against host workers (Boomsma and Nash 2008). In the host, selection is expected to favour counter-defences preventing the acquisition of chemical cues by parasitic myrmecophiles. Selection operates differently when a myrmecophile biosynthesizes chemical cues ("innate chemical mimicry" sensu chapter 4), because the origin of mimetic cues and model cues is different. This allows coevolutionary arms races to shape the degree of mimicry as well as the discrimination ability of ants.

Previous studies revealed that the silverfish and the spider showed specific behaviors to sustain physical contact to the host, e.g. they rubbed intensely against host workers (Witte et al. 2009). Thus, I expected them to acquire their mimetic compounds from the host rather than biosynthesing them. Under the assumption of an acquisition of mimetic CHCs from the host, the quantity of mimetic compounds is expected to decrease when myrmecophiles are isolated from their host. Accordingly, I isolated silverfish and spider individuals for several days and compared the concentration of CHCs (quantity of compounds per body surface) between isolated and non-isolated (unmanipulated) individuals (Chapter 2 and 3). The latter had host contact prior to chemical extractions. Additionally, the acquisition of host compounds was investigated by evaluating the transfer of a stable-isotope labelled hydrocarbon from the cuticle of host ants to the cuticle of myrmecophiles. Since both myrmecophiles were expected to acquire mimetic CHCs from their host, I hypothesized that both the spider and the silverfish will lose mimetic CHCs in the isolation experiment and that they will acquire the CHC label through physical contact with their host in the chemical-labelling experiment.

Although numerous studies have already described social insect parasites which apparently show surface chemicals resembling those of their hosts (Bagnères and Lorenzi

2010), the benefit of chemical mimicry has rarely been tested. As a consequence, most studies dealing with chemical mimicry remain descriptive. A chemical resemblance does not necessarily mean that the host is deceived by a mimic or that the mimic gains benefits through chemical resemblance. Mimicry in the strict sense only occurs when both of these circumstances are true (see chapter 4). Accordingly, specific bioassays are necessary to demonstrate whether chemical mimicry affects the behaviour of the host in a way that is beneficial for the mimic (Allan et al. 2002; Nash et al. 2008). I predicted that a good match of host and parasite chemical cues is a proximate mechanism protecting myrmecophiles from ant attacks, and consequently facilitates their social integration. Conversely, myrmecophiles with a poor chemical resemblance to the host should be treated more aggressively. To test these predictions, I investigated the silverfish and the spiders' dependency on chemical resemblance by performing aggression tests with individuals isolated from their hosts for extended periods. These individuals should then show lower chemical host resemblance and elicit higher aggression from the host compared to non-isolated (unmanipulated) individuals (Chapter 2 and 3). Table I summarizes the different research approaches and working hypotheses.

Table 1. Different approaches to studying the ultimate and proximate causes of different levels of integration among myrmecophiles and the underlying working hypotheses.

Research approach	Research topic	Hypotheses	Study species	Integrated[a]
Ultimate mechanisms (Chapter 1)	Interdependency of parasite impact and host defence	More costly myrmecophiles are attacked more frequently. Accordingly, they achieve lower integration levels.	Five staphylinid beetles: *Maschwitzia ulrichi*	No.
			Witteia dentilabrum	No.
			Parawroughtonilla hirsutus	Yes.
			Leptogenonia roslii	Yes.
			Togpelenys gigantea	Yes.
Proximate mechanisms (Chapter 2 and 3)	Origin of mimetic compounds[b]	The two studied myrmecophiles acquire mimetic compounds from the host.	Silverfish: *Malayatelura ponerophila*	Yes.
	Accuracy in chemical mimicry facilitates integration	Myrmecophiles showing a lower accuracy in chemical mimicry are attacked more often and, thus, achieve lower levels of integration.	Spider: *Gamasomorpha maschwitzi*	Yes.

[a] Preliminary studies assessed which species are integrated (not aggressed by ants, found inside the host nest) and which are non-integrated (aggressed by ants, found outside the nest)

[b] This topic was not studied to explain different levels of integration, instead it addressed the question how chemical mimicry is achieved.

Chapter 1

Differential host defense against multiple parasites in ants

Christoph von Beeren, Munetoshi Maruyama, Rosli Hashim and Volker Witte

The staphylinid beetle Maschwitzia ulrichi *preyed on the larvae of its host* Leptogenys distinguenda.
© *Christoph von Beeren*

2011

♦

von Beeren C, Maruyama M, Hashim R and Witte V (2011). Differential host defense against multiple parasites in ants. *Evolutionary Ecology*, 25:259-276.

Evol Ecol (2011) 25:259–276
DOI 10.1007/s10682-010-9420-3

ORIGINAL PAPER

Differential host defense against multiple parasites in ants

Christoph von Beeren · Munetoshi Maruyama · Rosli Hashim · Volker Witte

Received: 20 April 2010 / Accepted: 17 August 2010 / Published online: 5 October 2010
© Springer Science+Business Media B.V. 2010

Abstract Host–parasite interactions are ideal systems for the study of coevolutionary processes. Although infections with multiple parasite species are presumably common in nature, most studies focus on the interactions of a single host and a single parasite. To the best of our knowledge, we present here the first study on the dependency of parasite virulence and host resistance in a multiple parasite system. We evaluated whether the strength of host defense depends on the potential fitness cost of parasites in a system of two Southeast Asian army ant hosts and five parasitic staphylinid beetle species. The potential fitness costs of the parasites were evaluated by their predation behavior on host larvae in isolation experiments. The host defense was assessed by the ants' aggressiveness towards parasitic beetle species in behavioral studies. We found clear differences among the beetle species in both host–parasite interactions. Particular beetle species attacked and killed the host larvae, while others did not. Importantly, the ants' aggressiveness was significantly elevated against predatory beetle species, while non-predatory beetle species received almost no aggression. As a consequence of this defensive behavior, less costly parasites are more likely to achieve high levels of integration in the ant society. We conclude that the selection pressure on the host to evolve counter-defenses is higher for costly parasites and, thus, a hierarchical host defense strategy has evolved that depends on the parasites' impact.

Keywords Parasitism · Coevolution · Myrmecophiles · Fitness impact · Staphylinidae

Electronic supplementary material The online version of this article (doi:10.1007/s10682-010-9420-3) contains supplementary material, which is available to authorized users.

C. von Beeren · V. Witte (✉)
Department Biologie II, Ludwig-Maximilians Universität München,
Großhaderner Str. 2, 82152 Planegg, Germany
e-mail: witte@bio.lmu.de

M. Maruyama
The Kyushu University Museum, Fukuoka 812–8581, Japan

R. Hashim
Institute of Biological Sciences, Faculty of Science Building,
University Malaya, 50603 Kuala Lumpur, Malaysia

 Springer

Introduction

Coevolution is considered to be one of the most important processes shaping biodiversity on earth (Thompson 2005). It is characterized by reciprocal genetic modification in interacting species driven by natural selection, and it can emerge from different types of intimate interactions. Depending on the type of interaction, selection pressures may differ, e.g. among antagonistic predator or parasite systems versus mutualistic systems (Thompson 1994). Nevertheless, coevolving organisms are expected to exert specific selection pressures on their partners, which, in turn, lead to counter-adaptations in the partner, resulting in evolutionary arms races (Dawkins and Krebs 1979). Evolutionary theory predicts that each species should evolve in a way that fitness is maximized, which can lead to a conflict of interest between interacting species, assuming the partners are not closely related to each other (Axelrod and Hamilton 1981; Bronstein 2001; Sachs et al. 2003). Conflicts of interest and coevolutionary arms races may become particularly apparent in antagonistic interactions of host–parasite systems, which were studied here.

Parasitism is generally one of the most successful life strategies known among eukaryotes (de Meeûs and Renaud 2002). A large number of studies have addressed the interactions between a single host and a single parasite species (for an overview see Moore 2002; Combes 2005). In nature, however, most host species are affected by multiple parasite species (Petney and Andrews 1998; Read and Taylor 2001; Martens and Schön 2000; Rutrecht and Brown 2008). Such multiplicity of infection (also referred to as "parasitic coinfections", "concomitant infections" or "polyparasitism"; Bordes and Morand 2009) raises the question of whether a hierarchy of defensive behaviors exists, which depends on the severity of the parasitic impact as well as on the cost of the host response (Moore 2002). Numerous theoretical studies deal with the evolutionary consequences of multiple infections, mainly of different micro-parasite strains (Bremermann and Pickering 1983; May and Nowak 1995; Van Baalen and Sabelis 1995; Frank 1996; Brown et al. 2002; Schjorring and Koella 2003; Alizon et al. 2009). Competition between strains is usually expected to increase rather than decrease parasite virulence. The number of experimental studies that observe multiple parasitism is increasing. They show that infection with multiple parasites can either increase or decrease the parasites' virulence and, thus, the impact on the host species (Turner and Chao 1999; Perlman and Jaenike 2001; Barker et al. 2002; Bandilla et al. 2006; Bell et al. 2006; Rumbaugh et al. 2009). Additionally, different aspects of host defense have been likewise investigated under multiple parasite infections (Clayton et al. 1999; Allander and Schmid-Hempel 2000; Møller and Rósza 2005; Bordes and Morand 2009). However, none of these studies compares the impact of specific parasites in a host to the strength of host defenses targeting those parasites. In the present study, we directly address the question of whether a directed defense exists against more costly parasites in a multiple parasite situation.

We studied social insect colonies, which serve as hosts to a large variety of different parasites, including viruses, bacteria, fungi, protozoa, nematodes, helminthes, mites and insects (Schmid-Hempel 1988; Boomsma et al. 2005). Many species of insects and other arthropods have developed parasitic relationships with ants, especially with army ants (Wasmann 1895; Hölldobler and Wilson 1990; Gotwald 1995). Different classifications have been suggested to describe the diverse lifestyles of ant guests (Wasmann 1886; Deboutteville 1948; Paulian 1948; Akre and Rettenmeyer 1966). We use the broad distinction here between "integrated species", which are incorporated into the host societies by their own and their hosts' behavior, and "non-integrated species", which attain no integration into the host society but are nevertheless well-adapted to the host (Kistner 1979).

Studies on Neotropical myrmecophilous staphylinid beetles of ecitonine army ants have shown that there is a great diversity of parasite–host interactions in this particular beetle family (Wasmann 1895; Seevers 1965; Akre and Rettenmeyer 1966). Through convergent evolution, there is an analogous system that is situated in the Old World tropics. This system involves staphylinid beetles that are associated with ants of the genus *Leptogenys* (Formicidae: Ponerinae) in the Indomalayan ecozone, especially with those species showing army ant behavior (Kistner 1975, 1989; Kistner et al. 2003, 2008). A hierarchy of defense behaviors has recently been found in preliminary observations in one of the focal species of the present study, the ponerine army ant *Leptogenys distinguenda*. This species harbors a great variety of different parasite species, including staphylinid beetles (Witte et al. 2008). The behavior of these ants towards parasites ranges from tolerating some species to attacking, expelling or killing others. Because the ants' aggressiveness is easy to evaluate and it is possible to determine the impact for at least some parasites, army ants and their diverse parasite fauna represent a suitable model system to study multiple parasite systems.

To reduce the influence of taxonomic constraints, we focus in this study only on multiple parasitic beetle species (Staphylinidae) occurring in two related host ants, *L. distinguenda* and *L. borneensis*. We hypothesize that the magnitude of host defense depends on the costs imposed by the parasite. Thus, we predict that (1) the defense of ants should be stronger against more harmful parasites, and consequently (2) parasites that impose low costs are more likely to attain high levels of integration into the ants' social system.

Materials and methods

Field sampling

A total of 11 months of field work was performed between August 2007 and September 2009 in a regenerated, secondary dipterocarp lowland rainforest at the Field Study Centre of the University of Malaya (Kuala Lumpur), which is located in Ulu Gombak, Malaysia (03°19.4796'N, 101°45.1630'E, altitude 230 m) and at the Institute of Biodiversity in Bukit Rengit, Malaysia (03°35.779'N, 102°10.814'E, altitude 72 m). Five parasitic beetle species (Coleoptera: Staphylinidae) associated with two ponerine ant species, *Leptogenys distinguenda* and *Leptogenys borneensis*, were studied. In Ulu Gombak, colonies of *L. distinguenda* were inhabited by two beetle species, *Maschwitzia ulrichi* (formerly *Trachydonia leptogenophila*; Kistner et al. 2008) and *Witteia dentilabrum* n. gen. & sp. (Maruyama et al. in press a). Colonies of *L. borneensis* were inhabited by two different beetle species: *Parawroughtonilla hirsutus* n. gen. & sp. and *Leptogenonia roslii* n. gen. & sp. (M. Maruyama et al. in press b). In Bukit Rengit we found an additional beetle species, *Togpelenys gigantea* (Kistner 1989), in a single *L. distinguenda* colony. Only a limited number of studies could be carried out with *T. gigantea* because we found only three individuals.

To improve readability, we refer to *M. ulrichi* and *W. dentilabrum* as non-integrated species (NIS) and *T. gigantea*, *P. hirsutus* and *L. roslii* as integrated species (IS). For further information on these distinctions see discussion.

Both host species can reach large colony sizes (up to 50,000 workers in *L. distinguenda*; up to 5,000–10,000 workers participate in swarm raids in *L. borneensis*), are nocturnal and exhibit characteristic army ant behavior by performing massive collective raids and frequent colony migrations (Maschwitz and Steghaus-Kovac 1991; Steghaus-Kovac 1994; Witte 2001; Witte and Maschwitz 2002; Kronauer 2009). We located the nests during the night by back-tracking the ants' raiding trails. The nests were then marked and checked every 30 min for colony migrations between 8 p.m. and 3 a.m. Since all of the studied beetle species take part in the ants' migrations (*L. distinguenda* colonies migrate on average every 1.5 nights), they could be detected and collected during these activities. We sampled *L. distinguenda* and *L. borneensis* colonies using aspirators to capture ant workers, ant pupae and ant larvae as well as parasitic staphylinid beetles. Each collection was performed simultaneously by at least two people. Since we observed all migrations from the beginning to the end and as the beetles are rather conspicuous, it can be presumed that virtually all beetles of each ant colony were captured.

Migration structure

To study how the beetles participate in host migrations, we observed 21 migrations of *L. distinguenda* and seven migrations of *L. borneensis*. We recorded whether the beetles occurred during the ant migration or after it was already finished.

Laboratory maintenance

Studies on the behavior of the beetles as well as studies on host defense (see below) were performed in the field station in Malaysia with 13 laboratory colony fragments (eight *L. distinguenda* fragments and five *L. borneensis* fragments). The nest fragments included 110–170 ant workers, 44–55 ant pupae, 22–30 callows (freshly hatched workers) and three to six clusters of ant larvae as well as all of the staphylinid beetles collected in the respective colonies. A transparent plastic container (20 cm × 14 cm × 1 cm) with a 1 cm wide entrance was used as nesting space. It was placed into a larger foraging arena (32 cm × 22 cm × 5 cm) filled with a moistened plaster floor. The nesting space was kept dark during day time by covering it with a carton sheet. The side walls of the foraging arena were treated with FLUON (Whitford GmbH) and the arena was covered to prevent workers and beetles from escaping. Small pieces of freshly killed crickets were placed daily in the food arena. All observations were carried out between 8:00 p.m. and 4:00 a.m. using weak ambient light which did not noticeably affect the behavior of the nocturnal ants and their myrmecophiles.

Preferred location of beetles in laboratory nests

The preferred locations of beetles in the laboratory nests were monitored by random scan-sampling during 8:00 p.m. and 4:00 a.m. on ten different days. The minimum time span between two scan-samplings was 1 h. The beetles' locations were categorized as follows: (1) waste disposal site (hiding place outside the nest), (2) folded piece of moistened filter paper (hiding place outside the nest), (3) free in the foraging arena, (4) furrows in the plaster (hiding place inside the nest) and (5) free in the nest interior. The waste disposal site consisted of dead ant workers, open pupae cocoons and prey remnants and was typically located in the corner of the foraging arena. The folded piece of paper was placed in the opposite corner of the foraging arena.

Parasite impact

To estimate the potential cost of a given beetle species to its host, we studied their predatory behavior. First, each beetle was isolated and starved for 24 h in a small plastic box (5 cm × 4 cm × 4 cm). Then, one larva of the corresponding host species was offered. After another 24 h the larval survival was checked under a stereomicroscope by visual inspection and by gentle stimulation with a thin needle. Living larvae always reacted with noticeable movements. There were two possible outcomes of the experiment: (1) larva alive or (2) larva dead. As a control, we kept larvae isolated without beetles for 24 h and determined their survival rate. Additionally, we observed whether the beetles preyed upon ant larvae in laboratory nests during behavioral and integration studies (see below). Each beetle individual was tested up to six times at maximum. Repeated observations were considered in the statistical analysis (see below). After isolation with a larva, individuals were starved again for 24 h before a new larva was offered.

Host defense

In order to investigate the defensive behavior of the hosts, we quantified the level of host aggression against the parasites by performing a contact study in laboratory nests. For this purpose, we observed the interactions of one focal beetle in 50 consecutive encounters with host ant workers. Because colony sizes consisted of 110–170 workers, repeated interactions with the same individuals were possible. However, since we focused on colony-level defense, and since task allocation naturally occurs in social insects, repeated actions do not affect our interpretation. We defined different interaction categories (see Table 1). The waste disposal site and the moistened piece of paper were removed during this study in order to increase the likelihood of encounters.

Table 1 Interactions between host ants and beetles

Interaction	Definition	Category
Ignore	An ant worker touches the beetle with its antennae and continues without any sign of behavioral modification	Peaceful
Groom	An ant grooms the beetle with its mouthparts	Peaceful
Avoid	When an ant approaches, the beetle avoids contact by escaping	Neutral
Unnoticed	An ant comes into and perhaps stays in contact with a beetle, but not with its antennae; the ant does not modify its behavior	Neutral
Antennate	An ant remains in contact with the beetle and touches the beetle's body repeatedly with its antennae	Neutral
Appeasement	The beetle lifts up its abdomen tip, obviously appeasing ant workers (most likely by the release of chemicals from its abdominal gland)	Neutral
Chase	An ant touches the beetle with its antennae and quickly lunges in its direction	Aggressive
Snap	An ant touches the beetle with its antennae and snaps with its mandibles in its direction	Aggressive
Sting	An ant touches the beetle with its antennae, lunges forward and bends its gaster in the opponent's direction. The attempt does not need to be successful	Aggressive

For each beetle, interactions were recorded over 50 encounters to determine the level of host aggression

An aggression index (AI) was calculated for each individual from the observed interactions in order to quantify the level of aggression towards individuals. The various categories (peaceful, neutral and aggressive; Table 1) only describe the ants' reaction during encounters, thereby disregarding the beetle's behavior. The interaction "ignore" was defined as a peaceful behavior because the ants did not react, even though they had the chance to recognize and thereby attack the beetles. In addition, we defined "groom" as a peaceful behavior in ants. A prolonged inspection through "antennation" often occurred between workers from different colonies but not between nestmates (unpublished data). Interestingly, workers from different colonies were not attacked, as is typically the case for most other ant species. Instead, they were intensively groomed and afterwards they achieved full integration. Therefore, we define the interaction neither as peaceful nor as aggressive, but as neutral. In the categories "unnoticed" and "avoid", the ants had low chances to recognize the beetles, and consequently we defined them as neutral interactions. "Appeasement" is the beetles' reaction to prevent ant aggression and as such, it is not adequate to deduce actual aggression of the host. Thus, we defined it as a neutral behavior. The aggression index was calculated with the help of the described categorizations in the following way:

$$AI = \frac{(\text{number of aggressive interactions} - \text{number of peaceful interactions})}{\text{total number of interactions}}$$

Accordingly, the aggression index is positive if more interactions were aggressive (maximum = 1), zero if interactions were equally aggressive and peaceful, and negative if more interactions were peaceful (maximum = -1). The aggression index value was set to one if the beetle was captured by the ants, which only occurred once during this study.

Behavior of beetles

To study the behavior of beetles in laboratory nests, we quantified the occurrence of different behavioral patterns (Table 2) during time spans of 10 min. Longer lasting behaviors were recounted every minute, e.g. in the case that the beetle was hiding for a longer period of time. Other behavioral categories could not always be recorded during the hiding behavior as the beetles were not fully visible. Each individual beetle was observed over a period of 2 days after collecting it in the field. All individuals of the species *M. ulrichi* and *W. dentilabrum* found in *L. distinguenda* colonies were observed three times at most. Individuals of the species *P. hirsutus* and *L. roslii* from *L. borneensis* colonies were observed at maximum five times per individual due to their rareness. The number of observations for each beetle species and the number of individuals tested is given in Supplementary Tables 3 and 4 of the supplementary material. Repeated observations were considered in the statistical analysis (see below). During the observations of each individual beetle, we captured and separated all the other beetles to avoid confusion between the individuals.

Data analysis

Data were evaluated with the software PRIMER 6 (version 6.1.11, Primer-E Ltd., Ivybridge, UK). The results of the behavioral and integration studies as well as the preferred locations of beetles were evaluated by an analysis of similarity (ANOSIM) with 999 permutations on Euclidean distances using a 2-factor nested design (individuals nested within species). The data were transformed ($\log (x + 1)$) where necessary to reduce the

Table 2 Behavioral patterns of beetles at the host nests

Behavior of beetles	Definition
Contact with ant	Staphylinid beetle is in direct physical contact with an ant for longer than 2 s (either showing active behavior such as rubbing or grooming, or passive behavior, i.e. resting on top, below or besides the ant)
Contact with brood	Staphylinid beetle is in direct physical contact with ant larvae or pupae
Hiding	Staphylinid beetle hides somewhere in the nest setup (e.g. in the waste disposal site) without interacting with the host
Feeding	Staphylinid beetle feeds on host prey items (crickets)
Self-grooming	Staphylinid beetle grooms itself with its legs or mouthparts

The beetles' behavior was observed for 10 min in artificial laboratory ant nests and all listed behavioral patterns were counted

effects of outliers. If fewer than 600 unique permutations were possible, the actual number of permutations is given in the text. The migration study and the study on the parasites impact were analyzed using an ANOSIM as described above, but the resemblance measure was simple matching of presence and absence data because the response variables were binomial. Non-metric multidimensional scaling (NMDS) was applied to visualize differences between species. Other figures were created with Microsoft Office Excel 2007 including the Excel add-in SSC-Stat (version 2.18, Statistical service centre of the University of Reading, Reading, UK).

Results

Field sampling

In Ulu Gombak, we found 141 individuals of the NIS *M. ulrichi* (range = 0–19 individuals/colony; median = 5) and 29 individuals of the NIS *W. dentilabrum* (range = 0–9 individuals/colony; median = 1) in 21 *L. distinguenda* colonies. *Witteia dentilabrum* was observed occasionally participating in ant raiding columns (four occasions in 35 observed raid columns), and because migrations always originated from previous raids, beetles might have reached a new nest site before the onset of the colony migration. Thus, we possibly missed some *W. dentilabrum* individuals.

In Bukit Rengit, we found eight individuals of *M. ulrichi* (NIS) and three individuals of *T. gigantea* (IS) in one *L. distinguenda* colony migration.

From seven *L. borneensis* colonies in Ulu Gombak, we collected 12 individuals of the IS *P. hirsutus* (range = 0–6 individuals/colony; median = 1) and five individuals of the IS *L. roslii* (range = 0–2 individuals/colony; median = 1).

Migration structure

The behavior of beetle species during migrations differed significantly from each other, because they occurred at different migration stages (ANOSIM: $R = 0.474$, $P < 0.001$). In *L. distinguenda* colonies, *M. ulrichi* ($N = 141$; NIS) always followed nest emigrations after the last migrating ants (Fig. 1A). *Witteia dentilabrum* ($N = 29$; NIS) also followed

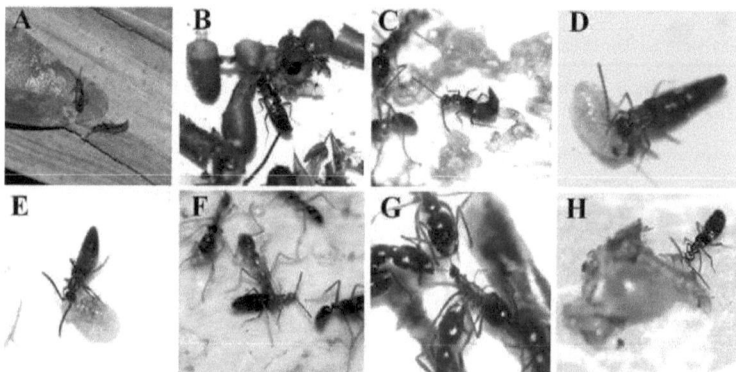

Fig. 1 Behavioral observations of staphylinid beetles. *Maschwitzia ulrichi* and *W. dentilabrum* followed the ants after the migration is finished presumably by perceiving the ant pheromone trail (**A** both *M. ulrichi*). Beetles of *L. distinguenda* hide for extended periods in the waste disposal sites of the ants (**B** *M. ulrichi*), whereas the beetles of *L. borneensis* stay in the host nest interior (**C** *P. hirsutus*).While *M. ulrichi* and *W. dentilabrum* prey on ant larvae in the feeding experiment (**D** *M. ulrichi*), the beetles from *L. borneensis* colonies sometimes lick the larva without inflicting harm to it (**E** *L. roslii*). *P. hirsutus* and *L. roslii* are treated peacefully by their host workers and have frequent contact with ant brood and workers (**C**, **F** *P. hirsutus*). The two beetle species *M. ulrichi* and *W. dentilabrum* are treated aggressively and are sometimes even caught by ants (**G**). All beetle species occasionally fed on pieces of dead crickets in laboratory nest fragments (**H** *P. hirsutus*)

afterwards ($N = 27$), or occasionally at the side of the migration column ($N = 2$), but never among the ant workers (21 observed migrations; Fig. 7a supplementary material). The NIS differed significantly in their preferred location (ANOSIM$_{M.\ ulrichi,\ W.\ dentilabrum}$: $R = 0.064$, $P = 0.038$).

In contrast, all three *T. gigantea* (IS) individuals followed the migration amidst the ant workers (one observed migration). Similar to the IS *T. gigantea*, the IS in *L. borneensis* colonies, *P. hirsutus* ($N = 12$) and *L. roslii* ($N = 5$) always moved among the migrating ants (seven observed migrations; Fig. 7b supplementary material). The position during migrations did not differ among the three IS (for all comparisons ANOSIM: $R = 0$, $P = 1$, unique permutations ≥ 56).

Most importantly, the IS differed significantly from the NIS in their occurrence during ant migrations (for all pairwise comparisons of IS and NIS: ANOSIM: $R \geq 0.799$, $P \leq 0.002$).

Preferred location of beetles in laboratory nests

The locations preferred by the different species in laboratory nest fragments differed significantly (ANOSIM: $R = 0.562$, $P < 0.001$; Fig. 2). However, the NIS *M. ulrichi* (N (individuals) $= 6$; N (observations) $= 32$) and *W. dentilabrum* (N (individuals) $= 6$; N (observations) $= 40$) did not differ in their preferred locations (ANOSIM: $R = -0.084$, $P = 0.781$, unique permutations $= 462$). Both NIS spent most of the time hiding in waste disposal sites (Fig. 1B).

 Springer

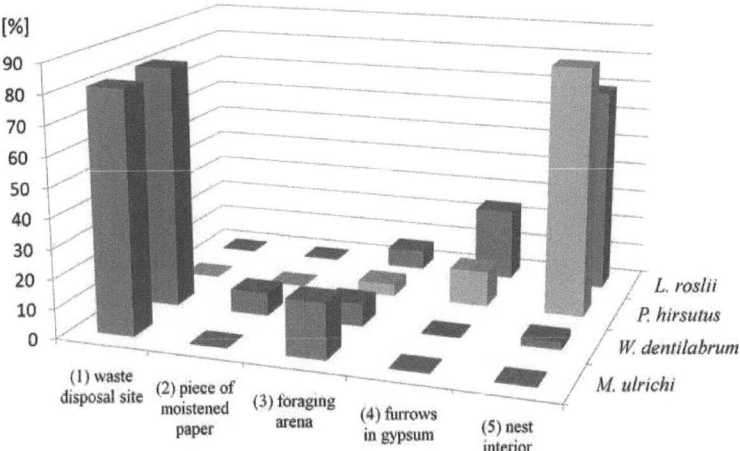

Fig. 2 Preferred locations of beetles in laboratory nests. *Maschwitzia ulrichi* and *W. dentilabrum* in *L. distinguenda* colonies preferentially stay in waste disposal sites, while both species in *L. borneensis* colonies remain mainly in the nest interior. Data were collected by randomly scan-sampling the locations of individuals in laboratory nests. *Abbreviations*: *NIS* non-integrated species, *IS* integrated species

In contrast, the IS *P. hirsutus* (N (individuals) = 3; N (observations) = 50) and *L. roslii* (N (individuals) = 3; N (observations) = 62) stayed mostly in the nest interior (Figs. 1C, 2). Their preferences did not differ significantly from each other (ANOSIM: $R = -0.185$, $P = 0.600$, unique permutations = 10), but the low number of permutations does not allow us to draw strong conclusions on this point. However, the preferred locations of the IS differed significantly from those of the NIS *M. ulrichi* and *W. dentilabrum* (ANOSIM: $R \geq 0.957$, $P = 0.012$; unique permutations = 84).

Togpenelys gigantea ($N = 3$; IS) stayed most of the time in the nest interior during the 6 h of observation time in laboratory nests, but we did not perform scan-sampling with this species.

Parasite impact

Most *L. distinguenda* larvae survived the 24 h isolation in the control experiments (larvae survived:larvae dead = 26:2; Fig. 3). In contrast, most of the larvae were killed when they were kept with individuals of the NIS, *M. ulrichi* ($N = 43$; larvae survived:larvae dead = 5:120) or *W. dentilabrum* ($N = 8$; larvae survived:larvae dead = 5:21). Accordingly, larval survival differed significantly from the control for both NIS species (ANOSIM$_{M.\ ulrichi,\ control}$: $R = 0.828$, $P < 0.001$; ANOSIM$_{W.\ dentilabrum,\ control}$: $R = 0.766$, $P < 0.001$). We repeatedly observed that the NIS species immediately attacked the larva, carried it around in their mandibles and fed on the nutritional haemolymph (Fig. 1D). However, we never observed any of the beetles preying on ant larvae in laboratory nests as the ants successfully expelled them from the nest interior by attack (see integration study below).

In three trials with three *T. gigantea* (IS) individuals, all larvae survived. However, the low sample size does not allow us to make strong inferences. We never found *T. gigantea*

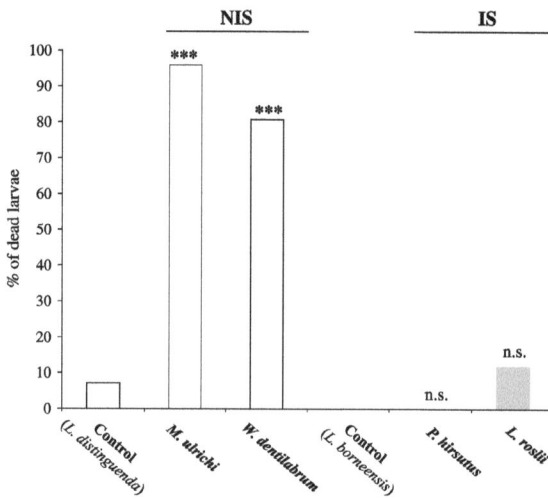

Fig. 3 Feeding experiment. In both control experiments, most larvae survived the isolation well. The two species *M. ulrichi* and *W. dentilabrum* are potential predators of ant larvae as they frequently killed the larvae. In contrast, we never observed an attack on host larvae from *P. hirsutus* or *L. roslii*, even though they were often in contact with the larvae in laboratory nests. Differences between the controls and the isolated larvae with beetles were evaluated using an ANOSIM (***$P < 0.001$). Data from beetles associated with the host *L. distinguenda* are shown by *white bars*, whereas the data concerning the beetles of *L. borneensis* are shaded in *gray*. *Abbreviations*: *NIS* non-integrated species; *IS* integrated species

(IS) preying on ant brood although they had frequent access to larvae in the laboratory nests (observation time circa 6 h).

All *L. borneensis* larvae survived the control experiment (larvae survived:larvae dead = 24:0). The IS *P. hirsutus* ($N = 5$; larvae survived:larvae dead = 12:0) and *L. roslii* ($N = 3$; larvae survived:larvae dead = 15:2) did not affect the survival of ant larvae compared to the control experiments (ANOSIM$_{P.\ hirsutus,\ control}$: $R = 0.000$, $P = 1$; ANOSIM$_{L.\ roslii,\ control}$: $R = 0.326$, $P = 0.125$). During at least 8 h of observation for each species in the behavioral and contact experiments, we never found any individual of these species attempting to feed on living host stages, although they frequently had contact with ant brood (Fig. 1E; see behavior study below).

Host defense

We found significant differences among beetle species in the contact study (ANOSIM: $R = 0.746$, $P < 0.001$). Three main groups can be distinguished (Fig. 4). The *M. ulrichi* and *W. dentilabrum* (NIS) group is mainly characterized by avoiding, being snapped and chased, the *P. hirsutus* and *L. roslii* (IS) group by remaining unnoticed and being ignored and the *T. gigantea* (NIS) group by ant grooming behavior.

The three IS remained more often unnoticed by their host (median$_{T.\ gigantea}$ = 19; median$_{P.\ hirsutus}$ = 27; median$_{L.\ roslii}$ = 28) than the two NIS (median$_{M.\ ulrichi}$ = 4; median$_{W.\ dentilabrum}$ = 4; see Table 3 in supplementary material for full detail).

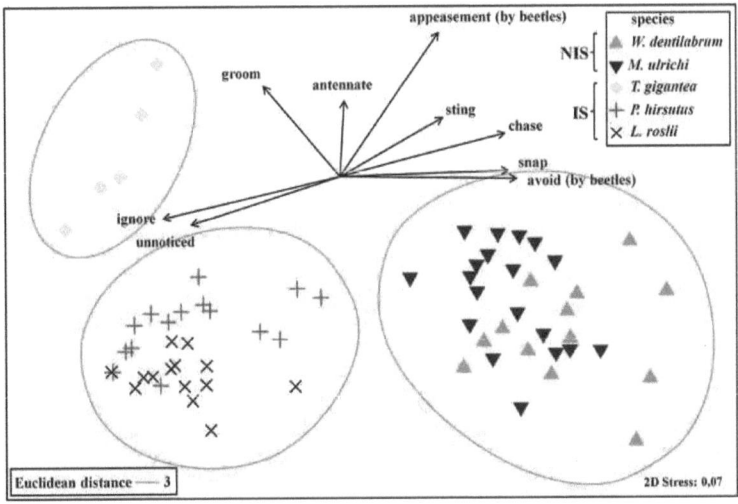

Fig. 4 Host defense. This nonmetric multidimensional scaling (*NMDS*) plot visualizes the differences among five beetle species in the host defense study. Each data point represents 50 encounters of an individual beetle with its host. Arrows visualize the contributions of behavioral categories to data separation, whereby the length indicates the importance. For clarity, the origin of arrows is not centered in the plot. 'Stress' is a quality measure of the NMDS. Distance = Euclidean distance. This resemblance measure can range from zero (=identical) to infinity. The maximum distance value for this data set is 7.2. *Abbreviations*: *NIS* non-integrated species, *IS* integrated species

Togpelenys gigantea is separated by NMDS from the two other IS, *P. hirsutus* and *L. roslii*, mainly because it was frequently groomed by the ants (median = 12; Fig. 4). We virtually never observed another beetle species being groomed by its host ant (see supplementary material, Table 3). Other interactions (antennation and appeasement) are less important for group separation and are therefore not evaluated further.

The aggression index of the different species differed significantly between two groups, one comprising the NIS *M. ulrichi* and *W. dentilabrum* and one the IS *T. gigantea*, *P. hirsutus* and *L. roslii* (ANOSIM: for all pairwise comparisons between species: $R \geq 0.983$, $P \leq 0.008$, number of permutations ≥ 120; Fig. 5). *Maschwitzia ulrichi* and *W. dentilabrum* (NIS) had an overall positive aggression index (median $(AI)_{M.\ ulrichi} = 0.20$; median $(AI)_{W.\ dentilabrum} = 0.28$) while *T. gigantea*, *P. hirsutus* and *L. roslii* (IS) had a negative aggression index (median $(AI)_{T.\ gigantea} = -0.48$; median $(AI)_{P.\ hirsutus} = -0.32$; median $(AI)_{L.\ roslii} = -0.36$; Fig. 5). The aggression index of *M. ulrichi* and *W. dentilabrum* (NIS) did not differ significantly (ANOSIM: $R = 0.095$; $P = 0.134$). They elicited a greater amount of aggressive interactions (e.g. chasing, snapping and stinging; Fig. 1G) and were rarely ignored or groomed by their respective workers in contrast to the IS *T. gigantea*, *P. hirsutus* and *L. roslii*. The aggression index of the IS did not differ significantly from each other (ANOSIM: $AI_{T.\ gigantea,\ P.\ hirsutus}$: $R = 0.278$, $P = 0.086$, number of permutations = 35; $AI_{T.\ gigantea,\ L.\ roslii}$: $R = 0.370$, $P = 0.10$, number of permutations = 10; $AI_{P.\ hirsutus,\ L.\ roslii}$: $R = -0.296$, $P = 1$, number of permutations = 35). For full information on all interactions see Table 3 in the supplementary material.

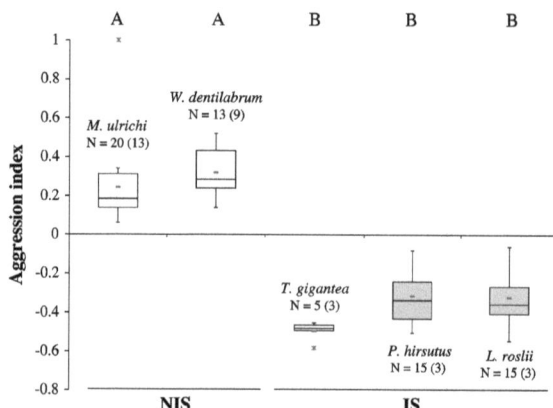

Fig. 5 Aggression index. The graph illustrates that the beetle species of *L. distinguenda* (*M. ulrichi* and *W. dentilabrum*) are treated with more aggression by their host than both beetle species of the ant *L. borneensis*. Only the species *T. gigantea* is integrated well in *L. distinguenda*. Different capital letters depict significant differences ($P < 0.05$) between groups evaluated by an ANOSIM. Data from beetles associated with the host *L. distinguenda* are white whereas the data concerning the beetles of *L. borneensis* are shaded in *gray*. *Abbreviations*: *N* number of observations (number of individuals), *NIS* non-integrated species, *IS* integrated species, — = mean, * = outlier

Behavior of beetles

We found significant differences among the beetle species across all behavioral categories (ANOSIM: $R = 0.695$; $P < 0.001$). Two groups, which match with host species and integration level, are clearly distinguishable (Fig. 6). The two groups (IS vs. NIS) are primarily separated by the behavioral categories hiding, contact with brood and contact with ant. The NIS *M. ulrichi* (median = 11) and *W. dentilabrum* (median = 8) were found hiding more frequently than the IS *L. roslii* (median = 1) and *P. hirsutus* (median = 0). Furthermore, the NIS *M. ulrichi* and *W. dentilabrum* rarely came into contact with their host ants (median for both species = 0), whereas the IS *P. hirsutus* (median = 35) and *L. roslii* (median = 50.5) had numerous contacts (Fig. 1F). Similar results were found for contacts with brood (NIS: median$_{M.\ ulrichi\ and\ W.\ dentilabrum}$ = 0, IS: median$_{P.\ hirsutus}$ = 22, median$_{L.\ roslii}$ = 18). Other behavioral categories (feeding and self-grooming) were less important for the separation of groups and are, hence, not further evaluated. Additional observations revealed that all beetle species fed occasionally on the host prey, i.e. crickets (Fig. 1H). Detailed information about specific behavioral patterns is reported in Table 4 of the supplementary material.

Discussion

Our study included two non-integrated beetle species, i.e. *M. ulrichi* and *W. dentilabrum*, which were frequently attacked by their host and mostly avoided direct contact with ants. Consequently, they were found outside of the nests and migrated separately from their host. In contrast, three beetle species were highly integrated, i.e. *T. gigantea*, *P. hirsutus*

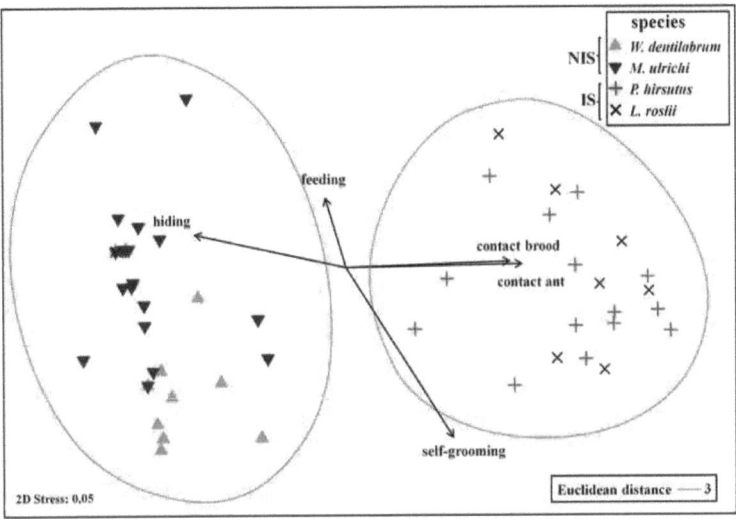

Fig. 6 Behavioral study. This nonmetric multidimensional scaling (*NMDS*) plot visualizes differences in behavior of four beetle species. Each data point is based on a 10 min observation of one beetle individual. Arrows visualize the contributions of behavioral categories to data separation, whereby the length indicates the importance. 'Stress' is a quality measure of the NMDS. Distance = Euclidean distance. This resemblance measure can range from zero (=identical) to infinity. The maximum distance value for this data set is 7.3

and *L. roslii*. These beetles were seldom attacked by their host and had frequent host contact. They lived in the center of the nests and migrated together with their hosts. Our central question in the following paragraphs is to explain the remarkable differences in the levels of social integration among the parasitic beetle species.

Proximate mechanisms: host aggression

The beetles' different levels of integration into the host societies can best be explained by the differential aggression these parasites receive. Non-integrated species were frequently attacked in contrast to integrated species, which interacted mainly peacefully with their hosts. Numerous aggressive interactions force intruders out of the center of the host colonies, where ant density is high and encounters are frequent. Under constant disturbance, the attacked species avoid host encounters and remain only in distant contact with their host. There is even the possibility for parasites to be captured and killed by the ants (Witte et al. 2009).

The recognition of alien intruders is a requirement for host defense to work. Nestmate recognition is based upon complex cuticular hydrocarbon profiles in social insects (Howard and Blomquist 2005; Hefetz 2007). Consequently, a likely explanation for the reduced aggression towards the integrated beetle species studied here is the failure of recognition either due to chemical mimicry or to chemical insignificance (Dettner and Liepert 1994; Lenoir et al. 2001). Indeed, the integrated species *P. hirsutus* and *L. roslii* show a higher degree of chemical resemblance than the non-integrated species *M. ulrichi* and *W. dentilabrum* (unpublished data). Nevertheless, other mechanisms such as behavioral adaptations (Witte et al. 2009) could exist so that this point deserves further investigation.

Ultimate mechanisms: impact on the host

Several parameters potentially shape evolutionary arms races between hosts and their parasites (Combes 2005). Generally, an adaptive response of one partner becomes more likely when stronger selection pressure is exerted by the other partner (Thompson 2005). More specifically, this can occur if parasites are highly virulent and reduce the fitness of their host considerably (Dawkins and Krebs 1979; Combes 2005). High virulence may result from several conditions, including the type of resources used (e.g. predation vs. kleptoparasitism), the parasites' body sizes, and their population densities (the latter two both influence the amount of resources used) (Witte et al. 2008). In the present example, size differences among beetle species are negligible (M. Maruyama et al. in press a) and the number of beetles per host colony (with a maximum of 19 individuals in a colony comprising thousands of workers) remains low compared to the host colony sizes. The predatory behavior, however, differs strongly among the beetle species, and this detrimental behavior clearly coincides with their level of social integration. Predation on ant larvae represents a potential fitness loss to the host, so that the selection pressure to evolve counter-defenses against predatory beetles is assumed to be higher. Consistent with this, the *L. distinguenda* host studied here defended itself successfully from detrimental intruders. Since laboratory and field data suggest that predatory beetles are successfully excluded from the nest interior of the host colonies, which typically houses the brood, a possible conclusion is that the host ants are leading the evolutionary arms races. Unlike the non-integrated species, the integrated beetle species are not predatory. Regarding their similar sizes and abundances (see above), kleptoparasitism on host diet imposes considerably lower costs to the host than predation on its brood. According to theory, selection for the evolution of counter-defenses is lower under such conditions (Dawkins and Krebs 1979; Combes 2005). Since there is no reason to assume that aggressive behavior towards the integrated species is generally more costly, our conclusion is that their higher social integration is likely a result of the lower costs they impose in terms of host fitness. This integration is beneficial to them because highly integrated species live in a stable and protected environment with reliable, high quality food resources (Hölldobler and Wilson 1990). Regarding these benefits, selection can possibly lead to reduced parasite virulence (see below).

Nevertheless, independent from the scenario described above, highly virulent parasites can still penetrate and live integrated inside of ant societies, if they are well adapted to exploit their host and are leading the arms race. The larvae of some lycanid butterfly species for example live in the nest interior of *Myrmica* (Formicidae: Myrmicinae) colonies, where they efficiently prey on ant larvae and thereby impose considerable damage to their host (Thomas and Wardlaw 1992). Nevertheless, the caterpillars appear to be sufficiently well integrated through sophisticated strategies to thrive inside the ant colonies (Akino et al. 1999; Barbero et al. 2009). Besides such extreme forms of parasitic exploitation, which may be stable due to frequency dependency or the dependence on additional partners (e.g. host plants; Pierce et al. 2002), we propose that in different associations the coevolutionary arms races are influenced by the fitness impact of parasites on their hosts, similarly as reported here.

Adjustment of host defense

Behavioral, mechanical or physiological host defenses can help in avoiding or reducing parasitism (Hart 1990; Boomsma et al. 2005; Delves et al. 2006). One possibility for coping with multiple infections is to direct the same type of defense equally against many or all parasites to lower the total cost and to maximize fitness accordingly. Common city doves (*Columba livia*), for example, limit the parasite load of two parasitizing feather lice species equally by efficient preening behavior (Clayton et al. 1999). In the systems studied here, frequent colony migrations represent a mechanism with the potential to reduce the overall parasite load, despite the fact that the symbionts have evolved different ways to follow their hosts (Witte et al. 2008).

A possibly less costly way to reduce parasite pressure under multiple infections is to direct defense preferentially against the most costly parasites, as suggested by Witte et al. (2008) for *L. distinguenda* myrmecophiles. Indeed, the present study gives additional evidence that *Leptogenys distinguenda* is able to detect and consequently direct their defense specifically against detrimental parasites.

Parasite pressure can also be reduced using several different defense mechanisms simultaneously. A study on birds suggests that in addition to preening behavior, the immune system could control the Amblyceran lice load by means of a T-cell mediated immune response (Møller and Rósza 2005). In our study, the ants' aggression is probably not the only defense (frequent colony migrations may serve as an additional counteraction), but it appears to be the most effective action against detrimental parasites and it could be used as a reliable measure of the hosts' defense against staphylinid beetles.

Why do different integration levels of staphylinids exist?

Hughes et al. (2008) argued that parasites of protected long-lived insect societies will tend to evolve reduced virulence. Additionally, they argued that large social insect colonies will have accumulated a higher diversity of low-cost parasites in comparison to the parasite diversity of small societies and nonsocial hosts. In this context, it is interesting that the integrated staphylinid beetles do not behave predatorily, because the predominant and plesiomorphic feeding habit in the subfamily Aleocharinae, which includes the studied species and most myrmecophilous Staphylinidae, is predation (MM; Thayer 2005). We therefore hypothesize that the integrated species could have lost their predatory lifestyle during the coevolution with their host and instead specialized on freshly killed prey items that are brought into the nest. The beetles benefit from this feeding preference, because the ants carry the costs of foraging and retrieving the food.

One hypothesis explaining the differences between integrated and non-integrated species might be competition among parasites. *Leptogenys distinguenda* and *L. borneensis* differ strongly in the composition of their parasite fauna. Among the studied taxa, only three symbiont species are known to occur in *L. borneensis* colonies in low numbers, i.e. the two staphylinids observed in this study plus one phorid fly species (Disney et al. 2009). In contrast, *L. distinguenda* colonies are parasitized by at least 15 different species (Witte et al. 2008; plus additional species under determination), some of which reach numbers of more than 1,000 individuals per colony (estimation of CvB). Several symbionts in *L. distinguenda* reach integration levels comparable to those observed for the integrated staphylinid beetles described in this study (Witte et al. 2008). Hence, it is possible that the niches for integrated species were already occupied and, thus, *M. ulrichi* and *W. dentilabrum* avoid competition for resources by occupying a different niche. Niche partitioning is a way to stabilize species diversity (Levine and HilleRisLambers 2009) and ant colonies offer many microhabitats that could be colonized by different species (Hölldobler and Wilson 1990). In another multiple parasite system, it was shown that 15 trematode species parasitizing the California hornsnail avoid competitive displacement by parasitizing different host tissues (Hechinger et al. 2009).

Convergent evolution of neotropical and Indo-Malayan staphylinid beetles

Wasmann (1895) noticed that the most frequently occurring ant guests are staphylinid beetles. He argued that this particular beetle family is preadapted for a myrmecophilous lifestyle. Although many staphylinid beetles of army ant colonies are described (Hölldobler and Wilson 1990), their behavior and exact interactions with their hosts often remain unknown. Interactions between staphylinids and ecitonine army ants in the Neotropics were studied intensively by Akre and Rettenmeyer (1966). In accordance with their observations, we found very similar differences in the social integration of staphylinid beetles associated with *Leptogenys* ants in the Indo-Malayan ecozone. Although the host species belong to different ant subfamilies (Neotropics: Ecitoninae; Indo-Malaysia: Ponerinae), they have independently evolved army ant behavior, i.e. they perform massive swarm raids and frequent colony migrations (Gotwald 1995; Kronauer 2009). Interestingly, myrmecophilous staphylinids apparently have likewise evolved convergent lifestyles, presumably due to similar adaptations to the army ant lifestyle.

Conclusion and future direction

Due to the fact that army ants are associated with various parasites, each imposing different costs, and that the ants' defensive behavior can be well quantified, they appear to be a suitable model to study the dependency of host defense on parasite impact in a multiple parasite system. Although some important aspects of parasitology still remain unknown in this army ant system, the results of the present study indicate that the hosts' defense and the impact of parasites are connected in that parasites imposing high costs are more likely to be fended off by the host. To the best of our knowledge, this is the first study that compares the strength of defense against multiple parasites dependent upon their individual impact. Future research will include the study of other parasite species of *L. distinguenda* colonies to evaluate whether the dependency between parasitic cost and host defense also holds for other taxonomic groups.

Acknowledgments We thank the behavioral ecology group at the LMU Munich and two anonymous reviewers for helpful comments on the manuscript. Many thanks are also due to Sofia Lizon à l'Allemand, Stefan Huber, Max Kölbl and Deborah Schweinfest for their assistance in the field. We are grateful for financial support from the DFG (Deutsche Forschungsgemeinschaft).

References

Akino T, Knapp JJ, Thomas JA, Elmes GW (1999) Chemical mimicry and host specificity in the butterfly *Maculinea rebeli*, a social parasite of *Myrmica* ant colonies. Proc R Soc Lond B 266:1419–1426
Akre RD, Rettenmeyer CW (1966) Behavior of Staphylinidae associated with army ants (Formicidae: Ecitoninae). J Kans Entomol Soc 39(4):745–782
Alizon S, Hurford A, Mideo N et al (2009) Virulence evolution and the trade-off hypothesis: history, current state of affairs and the future. J Evol Biol 22(2):245–259
Allander K, Schmid-Hempel P (2000) Immune defence reaction in bumble-bee workers after a previous challenge and parasitic coinfection. Funct Ecol 14(6):711–717
Axelrod R, Hamilton WD (1981) The evolution of cooperation. Science 211:1390–1396
Bandilla M, Valtonen ET, Suomalainen LR et al (2006) A link between ectoparasite infection and susceptibility to bacterial disease in rainbow trout. Int J Parasitol 36(9):987–991

Barbero F, Thomas JA, Bonelli S et al (2009) Queen ants make distinctive sounds that are mimicked by a butterfly social parasite. Science 323:782–785

Barker DE, Cone DK, Burt MDB (2002) *Trichodina murmanica* (Ciliophora) and *Gyrodactylus pleuronecti* (Monogenea) parasitizing hatchery-reared winter flounder, *Pseudopleuronectes americanus* (Walbaum): effects on host growth and assessment of parasite interaction. J Fish Dis 25(2):81–89

Bell AS, De Roode JC, Sim D et al (2006) Within-host competition in genetically diverse malaria infections: parasite virulence and competitive success. Evolution 60(7):1358–1371

Boomsma JJ, Schmid-Hempel P, Hughes WHO (2005) Life histories and parasite pressure across the major groups of social insects. In: Fellowes MDE, Holloway GJ, Rolff J (eds) Insect evolutionary ecology: proceedings of the royal entomological society's 22nd symposium. CABI, Wallingford, pp 139–176

Bordes F, Morand S (2009) Coevolution between multiple helminth infestations and basal immune investment in mammals: cumulative effects of polyparasitism? Parasitol Res 106:33–37

Bremermann HJ, Pickering J (1983) A game-theoretical model of parasite virulence. J Theor Biol 100(3):411–426

Bronstein JL (2001) The exploitation of mutualisms. Ecol Lett 4(3):277–287

Brown SP, Hochberg ME, Grenfell BT (2002) Does multiple infection select for raised virulence? Trends Microbiol 10(9):401–405

Clayton DH, Lee PL, Tompkins DM et al (1999) Reciprocal natural selection on host–parasite phenotypes. Am Nat 154(3):261–270

Combes C (2005) The art of being a parasite. The University of Chicago Press, Chicago

Dawkins R, Krebs JR (1979) Arms races between and within species. Proc R Soc Lond B 205:489–511

de Meeûs T, Renaud F (2002) Parasites within the new phylogeny of eukaryotes. Trends Parasitol 18:6

Deboutteville CD (1948) Recherches sur les Collemboles term-itophiles et myrmecophiles (ecologie, ethologie, systematique). Arch Zool Exptl Et Gen 85(5):261–425

Delves PJ, Martin SJ, Burton DR, Roitt IM (2006) Roitt's essential immunology. Blackwell publishing, Oxford

Dettner K, Liepert C (1994) Chemical mimicry and camouflage. Annu Rev Entomol 39:129–154

Disney RHL, Lizon à l'Allemand S, von Beeren C et al (2009) A new genus and new species of scuttle flies (Diptera: Phoridae) from colonies of ants (Hymenoptera: Formicidae) in Malaysia. Sociobiology 53(1):1–12

Frank SA (1996) Models of parasite virulence. Q Rev Biol 71(1):37–78

Gotwald WH Jr (1995) Army ants: the biology of social predation. Cornell University Press, Ithaca

Hart BL (1990) Behavioral adaptations to pathogens and parasites: five strategies. Neurosci Biobehav Rev 14:273–294

Hechinger RF, Lafferty KD, Mancini FT et al (2009) How large is the hand in the puppet? Ecological and evolutionary factors affecting body mass of 15 trematode parasitic castrators in their snail host. Evol Ecol 23(5):651–667

Hefetz A (2007) The evolution of hydrocarbon pheromone parsimony in ants (Hymenoptera: Formicidae)—interplay of colony odor uniformity and odor idiosyncrasy. A review. Myrmecol News 10:59–68

Hölldobler B, Wilson EO (1990) The ants. Harvard University Press, Cambridge

Howard RW, Blomquist GJ (2005) Ecological, behavioral and biochemical aspects of insect hydrocarbons. Annu Rev Entomol 50:371–393

Hughes DP, Pierce NE, Boomsma JJ (2008) Social insect symbionts: evolution in homeostatic fortresses. Trends Ecol Evol 23(12):672–677

Kistner DH (1975) Myrmecophilous Staphylinidae associated with *Leptogenys* Roger (Coleoptera; Hymenoptera, Formicidae). Sociobiology 1:1–19

Kistner DH (1979) Social and evolutionary significance of social insect symbionts. In: Hermann HR (ed) Social insects. Academic Press, New York, pp 339–413

Kistner DH (1989) New genera and species of Aleocharinae associated with ants of the genus *Leptogenys* and their relationships (Coleoptera: Staphylinidae; Hymenoptera: Formicidae). Sociobiology 15:299–323

Kistner DH, Witte V, Maschwitz U (2003) A new species of *Trachydonia* (Coleoptera: Staphylinidae, Aleocharinae) from Malaysia with some notes on its behavior as a guest of *Leptogenys* (Hymenoptera: Formicidae). Sociobiology 42:381–389

Kistner DH, von Beeren C, Witte V (2008) Redescription of the generitype of *Trachydonia* and a new host record for *Maschwitzia ulrichi* (Coleoptera: Staphylinidae). Sociobiology 52(3):497–524

Kronauer DJC (2009) Recent advances in army ant biology (Hymenoptera: Formicidae). Myrmecol News 12:51–65

Lenoir A, D'Ettorre P, Errard C et al (2001) Chemical ecology and social parasitism in ants. Annu Rev Entomol 46:573–599

Levine JM, HilleRisLambers J (2009) The importance of niches for the maintenance of species diversity. Nature 461(7261):254–257

Martens K, Schön I (2000) Parasites, predators and the Red Queen. Trends Ecol Evol 15(10):392–393

Maruyama M, von Beeren C, Rosli H (in press a) Myrmecophilous aleocharine rove beetles (Coleoptera: Staphylinidae) associated with *Leptogenys* Roger, 1861 (Hymenoptera: Formicidae) I. Review of three genera associated with *L. distinguenda* (Emery, 1887) and *L. mutabilis* (Smith,1861). Zookeys

Maruyama M, von Beeren C, Witte V (in press b) Aleocharine rove beetles (Coleoptera: Staphylinidae) associated with *Leptogenys* Roger, 1861 (Hymenoptera: Formicidae) II. Two new genera and two new species associated with *L. borneensis* Wheeler, 1919. Zookeys

Maschwitz U, Steghaus-Kovac S (1991) Individualismus versus Kooperation: gegensätzliche Jagd- und Rekrutierungsstrategien bei tropischen Ponerinen (Hymenoptera: Formicidae). Naturwissenschaften 78:103–113

May RM, Nowak MA (1995) Coinfection and the evolution of parasite virulence. Proc R Soc Lond B 261(1361):209–215

Møller AP, Rósza L (2005) Parasite biodiversity and host defenses: chewing lice and immune response of their avian hosts. Oecologia 142:169–176

Moore J (2002) Parasites and the behavior of animals. Oxford University Press, New York

Paulian R (1948) Observations sur les Coléoptères commensaux d'Anomma nigricans en Côte d"Ivoire. Ann Sci Nat Zool 10:79–102

Perlman SJ, Jaenike J (2001) Competitive interactions and persistence of two nematode species that parasitize Drosophila recens. Ecol Lett 4(6):577–584

Petney TN, Andrews RH (1998) Multiparasite communities in animals and humans: frequency, structure and pathogenic significance. Int J Parasitol 28(3):377–393

Pierce NE, Braby MF, Heath A et al (2002) The ecology and evolution of ant association in the Lycaenidae (Lepidoptera). Annu Rev Entomol 47:733–771

Read AF, Taylor LH (2001) The ecology of genetically diverse infections. Science 292(5519):1099–1102

Rumbaugh KP, Diggle SP, Watters CM et al (2009) Quorum sensing and the social evolution of bacterial virulence. Curr Biol 19(4):341–345

Rutrecht ST, Brown MJF (2008) The life-history impact and implications of multiple parasites for bumble bee queens. Int J Parasitol 38:799–808

Sachs J, Mueller UG, Wilcox TP et al (2003) The evolution of cooperation. Q Rev Biol 79:136–160

Schjorring S, Koella JC (2003) Sub-lethal effects of pathogens can lead to the evolution of lower virulence in multiple infections. Proc R Soc Lond B 270(1511):189–193

Schmid-Hempel P (1988) Parasites in social insects. Princeton University Press, Princeton

Seevers CH (1965) The systematics, evolution and zoogeography of staphylinid beetles associated with army ants (Coleoptera, Staphylinidae). Fieldiana Zool 47(2):139–351

Steghaus-Kovac S (1994) Wanderjäger im Regenwald-Lebensstrategien im Vergleich: Ökologie und Verhalten südostasiatischer Ameisenarten der Gattung *Leptogenys* (Hymenoptera: Formicidae: Ponerinae). Dissertation, Johann Wolfgang Goethe Universität, Frankfurt

Thayer MK (2005) 11. Staphylinoidea. 11.7. Staphylinidae Latreille, 1802. In: Kristensen NP, Beutel RG (eds) Handbook of zoology vol IV, part 2. Arthropoda: Insecta. De Gruyter, Berlin, pp 296–344

Thomas JA, Wardlaw JC (1992) The capacity of a *Myrmica* ant nest to support a predacious species of *Maculinea* butterfly. Oecologia 91:101–109

Thompson JN (1994) The coevolutionary process. University of Chicago Press, Chicago

Thompson JN (2005) The geographic mosaic of coevolution. University of Chicago Press, Chicago

Turner PE, Chao L (1999) Prisoner's dilemma in an RNA virus. Nature 398(6726):441–443

Van Baalen M, Sabelis MW (1995) The dynamics of multiple infection and the evolution of virulence. Am Nat 146(6):881–910

Wasmann E (1886) Über die Lebensweise einiger Ameisengäste. I Dtsch Entomol Z 30:49–66

Wasmann E (1895) Die Ameisen-und Termitengäste von Brasilien. I. Theil. Mit einem Anhange von Dr. August Forel. Verh K K Zool Bot Ges Wien 45:137–179

Witte V (2001) Organisation und Steuerung des Treiberameisenverhaltens bei südostasiatischen Ponerinen der Gattung Leptogenys. Dissertation, Johann Wolfgang Goethe Universität, Frankfurt

Witte V, Maschwitz U (2002) Coordination of raiding and emigration in the ponerine army ant *Leptogenys distinguenda* (Hymenoptera: Formicidae: Ponerinae): a signal analysis. J Insect Behav 15:195–217

Witte V, Leingärtner A, Sabaß L et al (2008) Symbiont microcosm in an ant society and the diversity of interspecific interactions. Anim Behav 76:1477–1486

Witte V, Foitzik S, Hashim R et al (2009) Fine tuning of social integration by two myrmecophiles of the ponerine army ant, *Leptogenys distinguenda*. J Chem Ecol 35:355–367

Supplementary Material

Title: Differential host defense against multiple parasites in ants

von Beeren C, Maruyama M, Rosli H and Witte V

Table 3. Behavioral actions of host workers towards beetles. The upper number in each array represents the sum and the lower number indicates the median of the corresponding interaction. Different capital letters depict significant differences (p < 0.05) among beetles for a given behavioral interaction evaluated by an ANOSIM. Data for beetles associated with the host *L. distinguenda* have a white background whereas the data concerning the beetles of *L. borneensis* are shaded in gray. Abbreviations: N = number of observations (number of individuals)

Interaction	*M. ulrichi* N = 19 (13)		*W. dentilabrum* N = 13 (7)		*T. gigantea* N = 5 (3)		*P. hirsutus* N = 15 (4)		*L. roslii* N = 15 (3)		Category
Ignored	43 / 3	A	14 / 0	B	63 / 13	C	253 / 17	C	264 / 18	C	Peaceful
Groom	0 / 0	A	0 / 0	A	65 / 12	B	2 / 0	A	0 / 0	A	Peaceful
Avoid	424 / 19	A	321 / 28	A	0 / 0	B	26 / 2	B	21 / 1	B	Neutral
Unnoticed	98 / 4	A	43 / 4	A	87 / 19	B	414 / 27	C	438 / 28	BC	Neutral
Antennate	41 / 2	AB	8 / 1	A	25 / 4	B	11 / 0	A	17 / 1	AB	Neutral
Appeasement	135 / 7	A	49 / 3	AB	17 / 3	A / B	47 / 3	AB	2 / 0	B	Neutral
Snap	90 / 4	AB	54 / 8	A	1 / 0	C	6 / 0	C	13 / 0	BC	Aggressive
Chase	148 / 7	A	121 / 10	A	0 / 0	B	8 / 0	B	7 / 0	B	Aggressive
Sting	10 / 0	A	22 / 0	A	0 / 0	A	0 / 0	A	0 / 0	A	Aggressive

Table 4. Behavior of beetles in or at the host nests. The upper number in each array indicates the sum and the lower number the median of a given behavior. Different capital letters depict significant differences (p < 0.05) among beetles for a given behavior evaluated by an ANOSIM. Data for beetles associated with *L. distinguenda* have a white background whereas data for beetles of *L. borneensis* are shaded in gray. Abbreviations: N = number of observations (number of individuals)

Behavior	M. ulrichi N = 25 (14)		W. dentilabrum N = 11 (6)		P. hirsutus N = 14 (6)		L. roslii N = 8 (3)	
Contact to worker	19 0	A	10 0	A	562 35	B	393 50,5	B
Contact to brood	0 0	A	0 0	A	292 22	B	143 18	B
Hiding	254 11	A	89 8	B	1 0	C	23 1	B
Feeding	43 0	A	0 0	A	10 0	A	1 0	A
Self grooming	42 0	A	126 11	B	85 5	B	38 3	AB

Figure 7. Typical examples of (a) *L. distinguenda* and (b) *L. borneensis* migration structures. The frequencies of workers heading towards the new nest were counted constantly for 90 s (blue lines), followed by a 90 s break. Additionally, we counted all staphylinid beetles present throughout the entire emigration. This means that even during the breaks, all beetles in the migration column were recorded. Thus, the beetles were assigned to progressive 3 min intervals. The NIS of *L. distinguenda* colonies (four *M. ulrichi* and two *W. dentilabrum* individuals) followed the ant trail after the ant migration was completed. In contrast, the IS of *L. borneensis* colonies (one *P. hirsutus* and one *L. roslii* individual) migrated within the ants' migration column. Abbreviations: N = number of individuals

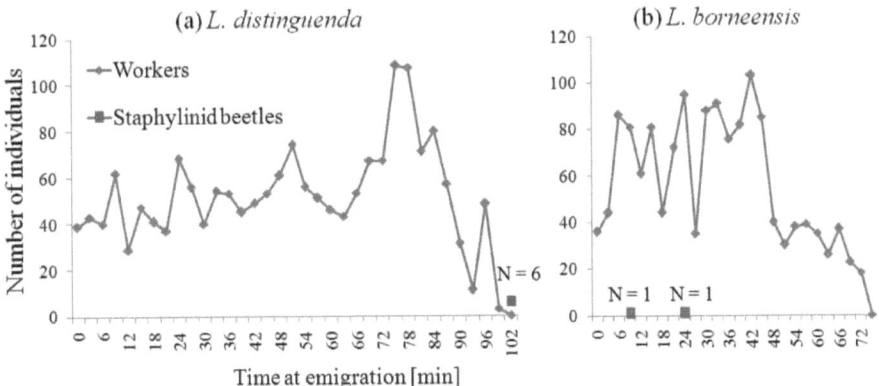

Chapter 2

Acquisition of chemical recognition cues facilitates integration into ant societies

Christoph von Beeren, Stefan Schulz, Rosli Hashim and Volker Witte

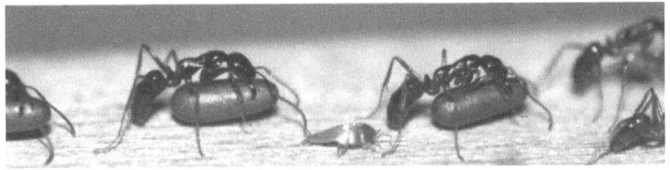

The kleptoparasitic silverfish Malayatelura ponerophila *is well integrated into host ant societies. It is one among many myrmecophiles that participate in the frequently occurring ant migrations of* L. distinguenda.
© *Christoph von Beeren*

2011

♦

von Beeren C, Schulz S, Hashim R and Witte V (2011). Acquisition of chemical recognition cues facilitates integration into ant societies. *BMC Ecology*, 11:30.

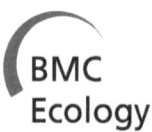

BMC Ecology

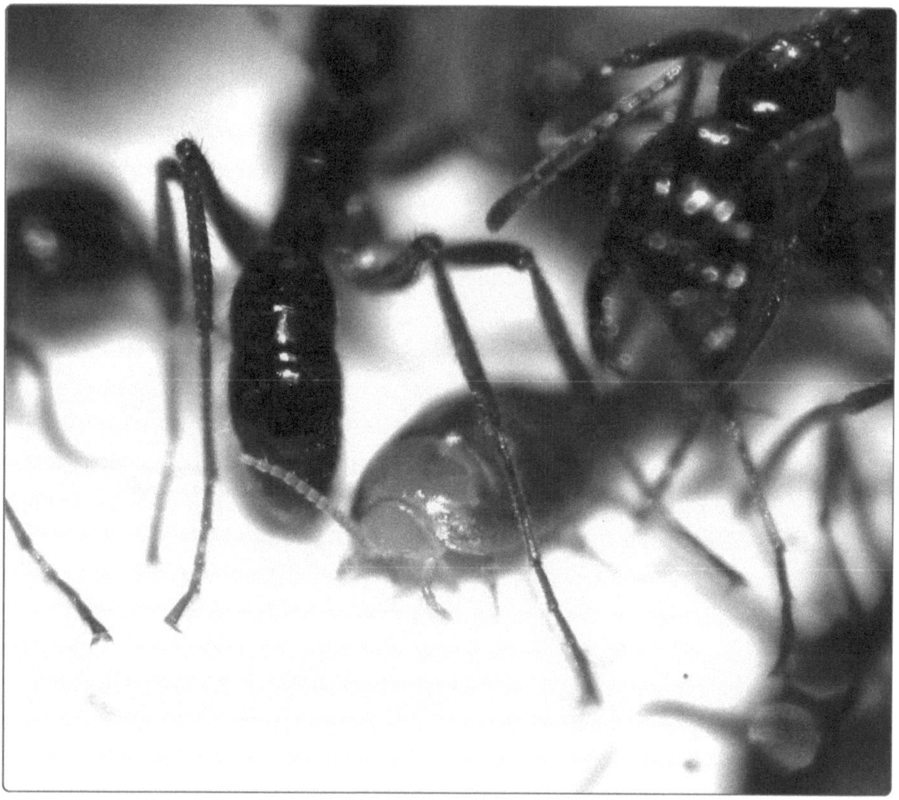

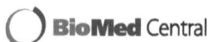

Acquisition of chemical recognition cues facilitates integration into ant societies

von Beeren *et al.*

BioMed Central

von Beeren *et al. BMC Ecology* 2011, **11**:30
http://www.biomedcentral.com/1472-6785/11/30 (1 December 2011)

RESEARCH ARTICLE

Acquisition of chemical recognition cues facilitates integration into ant societies

Christoph von Beeren[1], Stefan Schulz[2], Rosli Hashim[3] and Volker Witte[1*]

Abstract

Background: Social insects maintain the integrity of their societies by discriminating between colony members and foreigners through cuticular hydrocarbon (CHC) signatures. Nevertheless, parasites frequently get access to social resources, for example through mimicry of host CHCs among other mechanisms. The origin of mimetic compounds, however, remains unknown in the majority of studies (biosynthesis vs. acquisition). Additionally, direct evidence is scarce that chemical mimicry is indeed beneficial to the parasites (e.g., by improving social acceptance).

Results: In the present study we demonstrated that the kleptoparasitic silverfish *Malayatelura ponerophila* most likely acquires CHCs directly from its host ant *Leptogenys distinguenda* by evaluating the transfer of a stable-isotope label from the cuticle of workers to the silverfish. In a second experiment, we prevented CHC pilfering by separating silverfish from their host for six or nine days. Chemical host resemblance as well as aggressive rejection behaviour by host ants was then quantified for unmanipulated and previously separated individuals. Separated individuals showed reduced chemical host resemblance and they received significantly more aggressive rejection behaviour than unmanipulated individuals.

Conclusion: Our study clarifies the mechanism of chemical mimicry in a social insect parasite in great detail. It shows empirically for the first time that social insect parasites are able to acquire CHCs from their host. Furthermore, it demonstrates that the accuracy of chemical mimicry can be crucial for social insect parasites by enhancing social acceptance and, thus, allowing successful exploitation. We discuss the results in the light of coevolutionary arms races between parasites and hosts.

Background

Host-parasite interactions are often regarded as coevolutionary "arms races" in which a host and a parasite species exert reciprocal selection pressures on one another over long periods of time [1]. Under coevolution, parasite species adapt towards encountering a host and exploiting it successfully, whereas host species in turn adapt towards an avoidance of parasite encounters or a successful defence against them [2]. Accordingly, hosts have evolved a great variety of defence mechanisms to prevent all sorts of exploitative attacks [3,4].

Since social insects are widespread and extraordinarily abundant [5] they are subject to exploitation. As a consequence, they have evolved sophisticated recognition systems to protect their colonies, their brood, and their

* Correspondence: witte@bio.lmu.de
[1]Department of Biology II, Ludwig Maximilian University Munich, Großhaderner Straße 2, Planegg-Martinsried, 82152, Germany
Full list of author information is available at the end of the article

resources from competitors, predators and parasites, thereby maintaining the integrity of their societies [6]. The recognition of group members in social insects is mainly based on chemical cues [7-9]. Individuals compare the chemical cues expressed by a counterpart with an internal template, which is the chemical signature expected in all members of the society [10]. Complex blends of cuticular hydrocarbons (CHCs) seem to comprise all essential information necessary for nestmate recognition in ants, wasps and termites [11]. Due to effective recognition systems, invaders are frequently recognized, attacked, expelled or even killed by social insect workers.

Nevertheless, a multitude of organisms, particularly invertebrates, are known to exploit social insect societies [12-14], for example, by preying directly on the host, by stealing their food, or merely by inhabiting a well-protected habitat with a stable microclimate [6]. Many of these organisms, commonly known as myrmecophiles, are more or less permanently associated with ant colonies [15].

© 2011 von Beeren et al; licensee BioMed Central Ltd. This is an Open Access article distributed under the terms of the Creative Commons Attribution License (http://creativecommons.org/licenses/by/2.0), which permits unrestricted use, distribution, and reproduction in any medium, provided the original work is properly cited.

However, tight associations with ants require specific adaptations, that is, intruders must be able to invade host colonies and maintain contact without being expelled or killed. Some species not only manage to invade ant societies successfully, they also remain permanently integrated [6,16,17] in a way that the hosts behave amicably to the intruders as if they were part of their society [18].

A range of specific strategies exist to penetrate ant societies, and eventually to remain permanently integrated, including chemical, acoustic, morphological and behavioural adaptations [6,10,19-22]. Chemical strategies are particularly widespread among myrmecophiles, most likely because ants rely strongly on chemical communication [5,23]. Several chemical strategies have been described, such as chemical mimicry (chemical resemblance of another species), chemical camouflage/crypsis (avoiding detection through expression of uninteresting or background cues), chemical insignificance (suppression of chemical recognition cues) or the use of ant deterrent/attractant chemicals [10,19]. Pretending to be a member of the colony by mimicking the ants' CHCs (chemical mimicry) is among the most frequent chemical strategies among myrmecophiles [10,23]. Although another definition of chemical mimicry exists [24], we use this term consistently with its original biological definition according to Dettner & Liepert [19], irrespective of the mechanism through which these mimetic compounds are acquired. Nevertheless, to include this information, we consider chemical mimicry to be either innate (biosynthesis of compounds), acquired (adoption of compounds), or mixed.

Regarding the origins of mimetic compounds, we assume that the acquisition as well as the innate production of ant CHCs may be associated with costs for the myrmecophiles. As expected from a trade-off model, such costs must be balanced by a benefit of performing chemical mimicry [25], and this has rarely been tested empirically. In numerous cases, chemical mimicry presumably works through acquisition of host odours through physical contact rather than through biosynthesis [19,23]. However, the origin of mimetic compounds remains unclear in the majority of cases. As in many other examples, the kleptoparasitic silverfish *Malayatelura ponerophila* (Zygentoma, Atelurinae; Figure 1) was found to resemble the CHC profiles of its Southeast Asian army ant host, *Leptogenys distinguenda*. A closer analysis suggested the acquisition of host cues through physical contact, as the silverfish was observed to interact frequently with its host through rubbing its surface on that of host ant workers [26]. Nevertheless, final proof was lacking so that the mechanism remained speculative and a biosynthesis of mimetic cues could not be completely ruled out, as is the case in most other examples.

The aims of the present study were twofold; 1) to clarify the underlying mechanism of chemical mimicry (innate vs. acquired), and 2) to test whether a good match of chemical host recognition cues is in fact beneficial to the mimic (e.g., by facilitating social integration). Two experimental approaches were used to address these questions. First, we marked host ant workers with a stable isotope-labelled hydrocarbon and

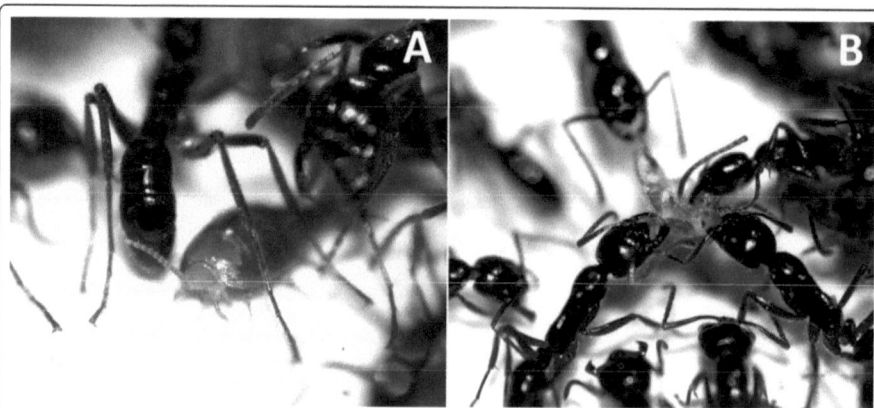

Figure 1 Interactions of *M. ponerophila* with host workers. (A) The silverfish is frequently found beneath host ant workers. Physical contact with the host ants allows the silverfish to acquire cuticular hydrocarbons, which are used by ants as recognition cues. (B) Life in ant colonies also entails a high risk for silverfish, as they are sometimes recognised, attacked, and killed by the host ants. © C. von Beeren.

monitored the transfer of this artificial label to the myrmecophilous silverfish. If *M. ponerophila* acquires CHCs from its host, we expected that the label would also accumulate on their cuticles. Second, we aimed to experimentally reduce the chemical host resemblance of silverfish individuals in order to study the effects on social integration. Therefore, silverfish were isolated from their host ants for six and nine days, respectively. Under the assumption of a behavioural acquisition of host recognition cues by the parasite, the silverfish were expected to lose host CHCs over time when isolated, resulting in reduced chemical similarity to the host. This assumption was checked by analysing cuticular chemical profiles of isolated and non-isolated silverfish and comparing them to their host. We expected that silverfish exhibiting reduced chemical host resemblance would be less socially integrated as a consequence of being increasingly recognized as alien. Hence, we tested through behavioural studies whether isolated silverfish (with reduced host similarity) were attacked more frequently than unmanipulated individuals.

Methods

(a) Field collection and animal maintenance

Animals were collected and observed at the Field Studies Centre of the University Malaya in Ulu Gombak, Malaysia (03°19.479' N, 101°45.163' E, altitude 230 m) and at the Institute of Biodiversity in Bukit Rengit, Malaysia (03° 35.779' N, 102°10.814' E, altitude 72 m). Ten months of field work were carried out in total between August 2008 and April 2011. We searched for *L. distinguenda* raiding trails during the night, tracked them back to their bivouac-like nests and subsequently checked every 30 min between 8 p.m. and 3 a.m. for the onset of a migration. The silverfish *Malayatelura ponerophila* [27] participate in migrations either by phoretic transport on host pupae or by following the ants' pheromone trail on their own [28]. The collected animals were either extracted directly for subsequent chemical analysis of CHCs, or they were maintained in artificial laboratory nests for various experiments (see below). Experimental colonies were assembled differently depending on the numbers of collected individuals and on the experimental protocol. Laboratory nests contained only members from one particular host colony (i.e., colonies were never mixed). If not described differently, nests consisted of a clear plastic container (20 × 14 × 1 cm), shaded with a plastic cover, and with a 1.5 cm wide entrance. The nest was placed in a larger foraging arena (32 × 25 × 9 cm) with a moistened plaster floor to maintain constant humidity. For isolation experiments animals were also kept separated from their home colonies in plastic containers (20 × 14 × 5 cm) equipped with a moist plaster floor. Animals in laboratory nests and those in isolation were fed every day with freshly killed crickets. Crickets are among the natural diet of the host ants [29] and the silverfish (personal observation). All behavioural studies were performed between 8:00 p.m. and 3:00 a.m. under dim scattered light since the focal animals are strictly nocturnal.

(b) Chemical transfer experiment

The chemical transfer experiment was carried out to test whether silverfish acquire CHCs from their host through physical contact. One hundred mature workers (collected from raids), 60 callows (newly hatched workers), approximately 40 larvae, 30 pupae, 21 silverfish and 10 non-myrmecophilous isopods (as a control) were kept in a nest constructed from natural materials (soil and leaf litter). The callows were treated with a stable isotope-labelled hydrocarbon (eicosane-d_{42}, C/D/N Isotopes Inc., Canada, Pointe-Claire). Callows were selected for the labelling treatment because they are less mobile [30] and less aggressive (CvB and VW, personal observation), and silverfish were found to interact preferentially with them [26]. Eicosane-d_{42} was used as a label because it has properties (chain length, molecular weight) similar to the CHCs that occur naturally on the host ants [26]. 200 μl of a saturated eicosane-d_{42} hexane solution were evaporated in a clean 20 ml glass vial so that the hydrocarbon fully covered the bottom of the vial as a solid film. The callows were then enclosed in the vial, which was moved gently for 30 min to transfer the labelled compound. Callows did not visibly suffer from this treatment. Labelled callows, untreated workers, silverfish, and control isopods were kept three days together in the laboratory nest and subsequently they were extracted for chemical analysis (details below). The isopods, collected from the natural habitat, were added to test whether eicosane-d_{42} transmits to animals in the nest environment that are not specifically in close contact with the host. Preliminary studies revealed that isopods were ignored by ants, which makes them ideal control animals. Ten additional isopods were directly labelled with eicosane-d_{42} as described above (labelled isopods) and extracted to verify that the isopod cuticle is able to adsorb the labelled compound.

(c) Isolation treatments

To manipulate the presence of host CHC profiles on silverfish (under the assumption of acquisition through physical contact) and to test for effects on behavioural interactions with the host ants, silverfish and host ants (as experimental and control groups, respectively) were separated from their home colonies and kept isolated for six (6d) or nine days (9d) (for sample sizes refer to table 1). Isolated and non-isolated individuals as well as some resident host workers were either extracted with hexane to analyse changes in their CHCs (colonies 1-3; see section (d) of the methods) or they were tested behaviourally in

Table 1 Overview of the number of silverfish individuals observed within each colony for the analysis of CHCs, the social acceptance experiment and the measurement of the silverfish' body surface area

Colony	Analysis of CHCs		Social acceptance experiment		Body surface area	
	No Isolation	Isolation	No Isolation	Isolation	No Isolation	Isolation
Colony 1	15	29**	-	-	15	27**
Colony 2	15	15**	-	-	15	15**
Colony 3	12	12*	-	-	10	12*
Colony 4	-	-	7	6*	-	-
Colony 5	-	-	18	24*	-	-
Colony 6	-	-	6	8*	-	-
			Accuracy of chemical mimicry			
Colony 7	21	14**	22	15**	21	14**

The number of silverfish individuals differs in some colonies between the multivariate analysis of all behavioural categories and the analysis of the aggression index as some individuals did not complete the standardised number of 50 ant contacts for the social acceptance experiment as some silverfish were seized by ant workers (Table 2). Abbreviations: * = six days isolated silverfish; ** = nine days isolated silverfish

their home colonies for social acceptance (colonies 4-6; see section (e) of methods), or both in combination (colony 7; see section (f) of the methods). The combined experiment (section (f) of the methods) was best suited for testing whether an individual's accuracy of chemical mimicry affects its level of social acceptance. The chemical (section (d) of the methods) and behavioural effects (section (e) of methods) were in addition studied independently, as an influence on the parasites' CHC signature through host contact during behavioural tests cannot be ruled out in the combined experiment.

A control experiment was performed with silverfish to determine whether the isolation treatment itself had an effect on their social acceptance, rather than changes in their CHC signature (e.g., due to physical suffering or adoption of additional compounds that originate from the experimental setup). As before, silverfish were isolated (for six days), and then one group was directly subjected to aggression tests and subsequently to chemical analyses, while the other group was allowed an additional 24 h contact with 50 host callows, before they were tested in the same way. The latter group thereby experienced the isolation treatment but was also given the chance to re-acquire host CHCs (silverfish isolation control; colony 8). Additional isolation control experiments were performed with adult ant workers collected from raids. The ant worker controls were conducted with three different colonies (colonies 9-11). These controls intended to test whether an isolation treatment similarly affects the expression of host worker CHC profiles (given the fact that they are able to biosynthesise the CHCs).

(d) Analysis of CHCs

Specimens were transferred individually into 2 ml vials with PTFE septa and extracted for 10 min in 200 µl hexane (HPLC grade, Sigma-Aldrich). After evaporation of the solvent, the CHCs were re-dissolved in 40 µl hexane containing an internal standard (methyl stearate, FLUKA Analytics, Sigma-Aldrich), and 20 µl were transferred into a 0.3 ml vial with limited volume insert (Chromacol, 03-FISV). Using an auto sampler (Agilent technologies, 7683 Series) 1 µl of each sample was injected into a gas chromatograph (Agilent Technologies 6890N) coupled to a mass spectrometer (Agilent Technologies GC 5975 MSD). Details on the methods can be found in [28].

Chemicals were identified by mass spectra and retention indices (RI), and peak areas were extracted using the software AMDIS (version 2.68) [31]. A target library of 109 compounds was created based on the compounds found on host ants and myrmecophiles [26]. As AMDIS uses the mass spectrum as well as the retention index to identify a substance, it has the advantage of reliably detecting compounds, even in low quantities. Structural alkene isomers were distinguished although double bond positions were not determined.

The absolute quantity of each compound was calculated using the internal standard (concentration = 20 ng/µl). The resulting total quantity of a sample was divided by the animals' surface area in square millimetres in order to standardise to a presumably perceivable concentration of chemicals by an ant's antennal contact and to control for size differences between animals. To calculate surface areas, the bodies of silverfish, workers and isopods were subdivided into geometrical areas and the relevant dimensions were measured using a stereomicroscope (Zeiss Stemi 2000-C) with a measuring eyepiece (see additional file 1: Calculation of animal surface areas). The surface area of silverfish was calculated for each individual separately because they varied considerably in size, while a median surface area was used for workers as well as for isopods. Specimens were stored in pure ethanol.

(e) Social acceptance experiments

The host's aggression toward individual silverfish or individual workers was quantified through a standardised contact study in laboratory nests. The nests contained

200 ant workers, which were collected from raids, because foraging workers behave more aggressively and are thus more likely to defend the colony [32].

Furthermore, the ants were given 1 h time to settle in the laboratory nest before starting the experiments because ants tend to behave more aggressively in familiar territory than in an unfamiliar setting [33]. Fifty consecutive encounters of a silverfish individual (or worker individual) and ants were then categorized according to table 2. Each individual was tested only once. However, repeated interactions with the same ant individuals were possible. Nevertheless, since we focused on a colony-level defence, which naturally includes task allocation, repeated interactions of the same workers do not affect our interpretations. An aggression index (AI) was calculated for each silverfish from the observed interactions as follows: AI = N_A/N_T with N_A = number of aggressive interactions and N_T = total number of interactions.

Some silverfish (N = 14) were seized by the ants before 50 encounters were completed. These individuals were removed to prevent their destruction so that they could be used for chemical analysis and body measurements. Although these individuals did not reach 50 host encounters, their AI was calculated as described above.

(f) Accuracy of chemical mimicry

To directly test the relation of chemical host resemblance to social acceptance for the same individuals, we combined the social acceptance study and the analysis of CHCs in one *L. distinguenda* colony (colony 7; table 1). For each silverfish individual host aggression was quantified first via social acceptance experiments (standardised contact study), and then its CHCs were extracted and subjected to chemical analysis.

(g) Data analysis

Chemical and behavioural data were evaluated with the software PRIMER 6 (version 6.1.12, Primer-E Ltd., Ivybridge, U.K.) with the PERMANOVA+ add-in (version 1.0.2) using a non-parametric permutational analysis of variance (PERMANOVA) with 9,999 permutations [34]. PERMANOVA models were based on Bray-Curtis similarities (as a semi-metric measure), either calculated from a single response variable (chemical similarity, CHC concentration, aggression index), or from numerous response variables (CHC profiles, behavioural interactions). Nonmetric multidimensional scaling (NMDS) was used to visualise multivariate data (PRIMER 6). Box plots were created from univariate data with the Microsoft Excel add-in SSC-Stat (version 2.18, Statistical service centre of the University of Reading, Reading, U.K.). Chi-square tests were accomplished using XLSTAT (Version 2010.3.06, Addinsoft, U.S.A.).

Chemical analysis

Since no silverfish-specific compounds were detected, the principle compounds that together constituted 99% to the chemical profiles of workers (N = 44) according to a similarity percentage analysis (SIMPER) on Bray-Curtis similarities were included in the statistical analysis of the CHC composition, the presence or absence of CHCs and the total CHC concentration and the chemical similarity (N = 32 compounds; see additional file 2: Table of compounds). To test whether the chemical similarity of silverfish to their host colony was influenced by isolation treatments, Bray-Curtis similarities to the average worker CHC profile of the respective host colony were used as a univariate response variable, and a PERMANOVA with a 2-factor nested design (colonies (random), days of isolation (fixed), nested in colony) was applied. No chemical worker profiles were available for colony 3. To test for additional differences in the quantity of CHCs, absolute concentrations (per surface area) were analysed in the same way.

Furthermore, multivariate approaches were used to analyse relative changes in CHC composition (Bray-Curtis similarities), and the presence or absence of compounds (simple matching). A PERMANOVA with a 2-factor

Table 2 Behavioural interactions between silverfish and ants and behavioural categories used for calculating the aggression index

Behaviour	Definition	Category
Ignored	An ant worker touches the silverfish once with its antennae and moves on without any sign of behavioural modification.	-
Groomed	An ant grooms the silverfish with its mouthparts. The silverfish remains in position.	-
Avoid	When an ant approaches, the silverfish avoids contact by quick escape.	-
Antennated	An ant touches a silverfish repeatedly with its antennae for longer than two seconds without displaying other behaviours.	-
Unnoticed	An ant comes into and perhaps stays in contact with a silverfish, but not with its antennae; the ant does not modify its behaviour.	-
Chased	An ant touches the silverfish with its antennae and quickly lunges in its direction.	Aggressive
Snapped	An ant touches the silverfish with its antennae and snaps with its mandibles in the direction of the silverfish.	Aggressive
Stung	An ant touches the silverfish with its antennae, lunges forward and bends its gaster in the direction of the opponent. The attempt is not necessarily successful.	Aggressive
Seized	An ant snapped at and subsequently seized a silverfish in its mandibles.	Aggressive

nested design as described above was applied for both analyses. Chromatograms of chemical profiles of host ants and silverfish can be found in an earlier article of one of the authors [26].

Behavioural analysis

Aggression indices of isolated vs. non-isolated individuals were compared using a PERMANOVA with a two-factor nested design as described above. The interactions of silverfish with their host ants were evaluated in a multivariate approach including all observed behaviours. These were standardised by the total number of interactions and a 2-factor nested design as described above was applied.

Results

(a) Chemical transfer experiment

Previously labelled callows still carried high concentrations of eicosane-d_{42} after the three-day experimental phase (median = 46.18 ng/mm^2; Figure 2). Interestingly, the concentration of eicosane-d_{42} did not differ between silverfish (median = 44.57 ng/mm^2) and callows (median = 46.18 ng/mm^2; PERMANOVA, P = 0.986), while lower concentrations were found on adult workers (median = 10.60 ng/mm^2; PERMANOVA, for both comparisons P < 0.001). Almost no eicosane-d_{42} was found on control isopods (median = 0 ng/mm^2), which consequently differed from labelled callows, workers and silverfish (PERMANOVA, for all comparisons P < 0.001). High quantities of eicosane-d_{42} on the labelled isopods (median = 100.13 ng/mm^2) demonstrated that their cuticle has the potential to adsorb the labelled CHC.

(b) Analysis of CHCs

Seventy compounds were detected on workers (N = 44). No silverfish-specific CHC was found (N = 133). The number of detected host compounds on silverfish decreased after isolation treatments (no isolation: $N_{compounds}$ = 28, $N_{silverfish}$ = 63; 6 days isolation: $N_{compounds}$ = 22, $N_{silverfish}$ = 12; 9 days isolation: $N_{compounds}$ = 23, $N_{silverfish}$ = 58). No compounds were detected on some of the specimens that had been isolated for 9 days (5 out of 58).

Non-isolated silverfish were chemically closer to their host workers than isolated individuals (PERMANOVA, for all colonies P ≤ 0.025; Figure 3). Significant differences between isolated and non-isolated silverfish were detected in the relative composition, the presence or absence, and in the total concentration of CHCs in three out of four different colonies (table 3). In colony 3 there was a trend that the compositions of CHCs differed between non-isolated and isolated silverfish (PERMANOVA, P = 0.064), whereas the presence or absence of CHCs and the CHC concentrations did not differ (PERMANOVA, P ≥ 0.134). For this colony we had the smallest sample size (see table 1).

Non-isolated silverfish carried higher total concentrations of host CHCs on their body (median = 55.23 ng/mm^2, N = 40) than isolated silverfish (median $_{6\ days}$ = 10.95 ng/mm^2, N = 12; median $_{9\ days}$ = 13.98 ng/mm^2, N = 42; PERMANOVA, P < 0.001; for within colony comparisons see additional file 3: Concentrations of

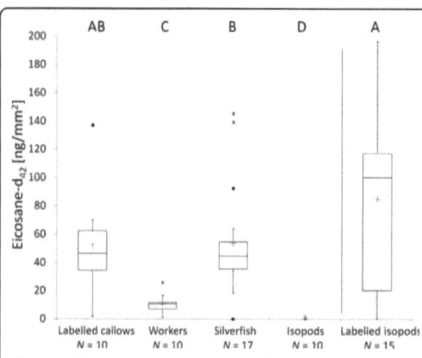

Figure 2 Concentrations of eicosane-d_{42} in the CHC transfer experiment. Different capital letters show significant differences (P < 0.05) between groups evaluated by PERMANOVA. Median (red cross = mean), quartiles (boxes), 10th and 90th percentiles (whiskers), and outliers (black square = outlier, asterisk = extreme point) are shown.

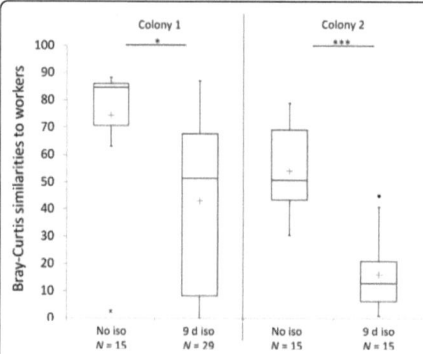

Figure 3 Chemical similarities of individual silverfish to the average chemical worker profile of their host colony ($N_{workers}$ ≥ 10). No chemical worker profiles were available for colony 3. Differences between isolated vs. non-isolated silverfish were evaluated by PERMANOVA (*P < 0.05, ***P < 0.001). Median (red cross = mean), quartiles (boxes) 10th and 90th percentiles (whiskers), and outliers (black square = outlier; asterisk = extreme point) are shown. Abbreviations: No iso = no isolation, d iso = days isolated.

Table 3 Comparison of non-isolated silverfish (0 d) and isolated silverfish (6 d or 9 d) regarding their CHC composition, presence or absence of CHCs and their total CHC concentration

Colony (Isolation of silverfish)	CHC composition	CHC presence/absence	CHC concentration
Colony 1 (0 d vs. 9 d)	0.015	0.001	0.001
Colony 2 (0 d vs. 9 d)	0.010	0.001	0.001
Colony 3 (0 d vs. 6 d)	0.064	0.801	0.134
Colony 7 (0 d vs. 9 d)	0.001	0.002	0.005

PERMANOVA P values are shown. For sample sizes see table 1. Abbreviations: d = days isolated

CHCs). Workers carried significantly higher concentrations than both silverfish groups (median = 106.23 ng/mm^2, N = 44; PERMANOVA, P < 0.001). Across all colonies the median concentration of every compound was lower after isolation.

(c) Social acceptance experiment

In all colonies, isolated silverfish were treated with higher aggression by host workers than non-isolated silverfish (PERMANOVA, for all comparisons P ≤ 0.004; Figure 4). The higher aggression toward isolated silverfish was also reflected in the frequency with which they were seized by workers. Only 4% of non-isolated silverfish were seized, while 26% of the six day isolated and 20% of nine-day isolated individuals were seized. All isolated silverfish were seized by workers in colony 6. The frequencies of seized and non-seized silverfish did not differ between six and nine days isolated silverfish (Chi square test: χ^2 = 0.232, df = 1, P = 0.630, N_1 = 38, N_2 = 15), but they differed

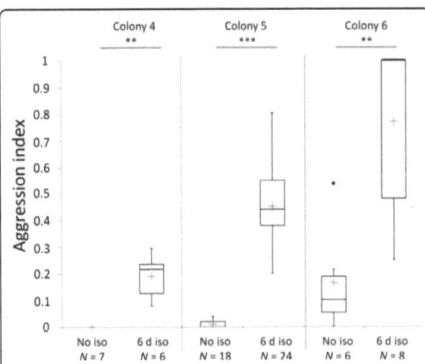

Figure 4 Host aggression in three different colonies toward non-isolated silverfish and silverfish that were isolated for six days. Differences between groups were evaluated by PERMANOVA (**P < 0.01, ***P < 0.001). Median (red cross = mean), quartiles (boxes), 10th and 90th percentiles (whiskers), and outliers (black square = outlier) are shown. Abbreviations: No iso = no isolation, d iso = days isolated.

significantly between non-isolated and isolated individuals (Chi square test: χ^2 = 11.851, df = 1, P = 0.001, N_1 = 53, N_2 = 53).

Considering the multivariate analysis of all behavioural interactions, we found significant differences in three of four colonies (colony 4, 5 and 7) between non-isolated and isolated silverfish (PERMANOVA, for all pair-wise comparisons P ≤ 0.012). Colony 6 was not evaluated because all of the isolated individuals were seized by worker ants and could not complete the standardised number of 50 ant contacts. For detailed information on behavioural interactions across all colonies see additional file 4: Behavioral interactions.

(d) Accuracy of chemical mimicry

In the experiment on the accuracy of chemical mimicry, the cuticular profile of isolated silverfish was also less similar to host workers (PERMANOVA, P < 0.001) and the same silverfish individuals received more aggression in contact studies than non-isolated individuals did (PERMANOVA, P ≤ 0.004; Figure 5). As in the experiments described above the total concentration of CHCs was lower in isolated silverfish (median = 4.51 ng/mm^2, N = 21; PERMANOVA, P < 0.001) than in non-isolated silverfish (median = 27.66 ng/mm^2, N = 21). Furthermore, non-isolated silverfish remained unnoticed more often and were ignored more frequently compared to isolated individuals (see additional files 4 and 5). Isolated silverfish were more frequently antennated by host workers, and they avoided host contact more often than non-isolated silverfish. Most importantly, ant workers chased and snapped at isolated silverfish more frequently than at non-isolated silverfish. There were no significant differences in the interactions "groomed" (PERMANOVA, P = 0.364) and "stung" (PERMANOVA, P = 0.365) between isolated and non-isolated silverfish (see additional file 4).

In the isolation control experiment silverfish that were first isolated for six days and then kept together with callows for 24 h showed greater chemical similarity (PERMANOVA, P < 0.001) and were treated less aggressively (PERMANOVA, P < 0.001) than individuals that were only isolated but had no secondary contact to the host (see additional file 6: Silverfish isolation control

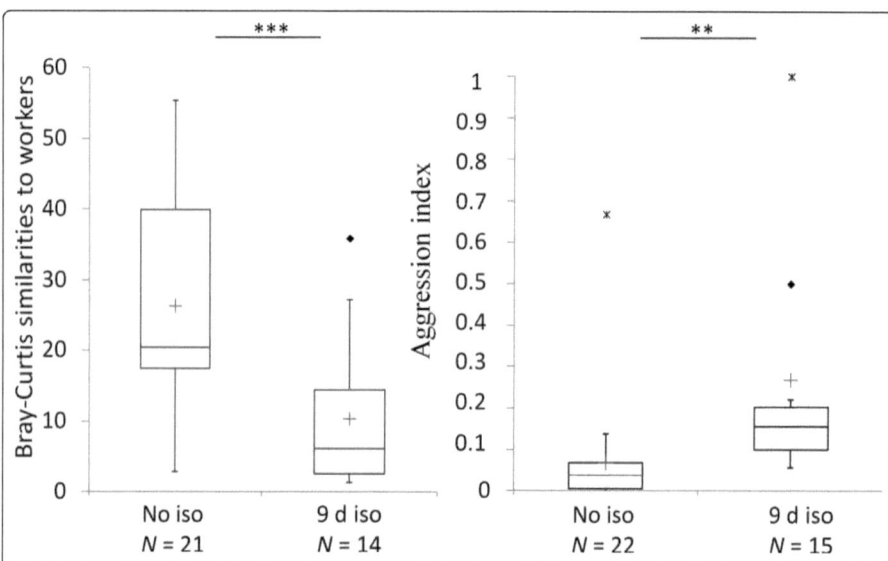

Figure 5 Chemical similarities of silverfish to the average chemical worker profile ($N_{workers}$ = 19; left) and host aggression toward the same individuals (right) in colony 7. Differences between isolated and non-isolated silverfish were evaluated by PERMANOVA (**P < 0.01, ***P < 0.001). Median (red cross = mean), quartiles (boxes), 10th and 90th percentiles (whiskers), and outliers (black square = outlier; asterisk = extreme point) are shown. Abbreviations: No iso = no isolation, d iso = days isolated.

experiment). In the host worker isolation control experiment, CHC concentration did not decrease after isolation, instead it increased in two colonies (additional file 7: Worker control experiment). Furthermore, aggressive behaviour did not increase (no isolation: three aggressive interactions from a total of 2282 recorded interactions; nine days isolation: five aggressive interactions from a total of 2777 recorded interactions).

Discussion

The present study sheds light on two important aspects of social insect parasitism. Our results strongly indicate that mimetic CHCs are acquired by a parasite from the cuticle of its host and that higher accuracy in mimicking host CHCs can be crucial for social exploitation due to the avoidance of aggressive rejection. In the following paragraphs we discuss in detail the integration mechanism of the parasitic silverfish M. ponerophila.

Origin of mimetic compounds (acquisition vs. biosynthesis)

The adoption of a stable isotope-labelled hydrocarbon from the cuticle of the host by the silverfish but not by control animals indicates that silverfish use a behavioural mechanism for acquiring mimetic CHCs, rather than innate biosynthesis. Eicosane-d_{42} has properties (chain length, molecular weight) similar to CHCs that occur naturally on the host ants, thus we conclude that the hosts' natural surface compounds are acquired by the same mechanism. Although we did not directly tested natural host CHCs, these compounds are most likely transferred in the same way. We cannot imagine a mechanism by which silverfish acquire selectively only particular compounds from the host cuticle. Furthermore, the mechanism of pilfering host CHCs (e.g., rubbing behaviour [26]; see also additional file 8: Video of M. ponerophila) appears to be very effective, as silverfish accumulated even higher concentrations of eicosane-d_{42} from the labelled callows than host workers did. In agreement with a behavioural adoption, the mimetic CHCs on silverfish decreased in isolation treatments quantitatively (total concentration) and qualitatively (relative abundance and presence or absence). Importantly, mimetic CHCs increased again after secondary contact of previously isolated silverfish with host ants. Taken together, the loss of mimetic cues after isolation and their reoccurrence after secondary host contact point strongly towards an effective behavioural acquisition of host

CHCs. Alternatively, these findings could be explained by a context-specific up- and down-regulation of CHC biosynthesis in the silverfish. However, due to the direct transfer of the labelled compound from host to silverfish, and due to evolutionary considerations that we explain below, this seems highly unlikely.

The exchange of surface compounds through physical contact (trophallaxis, allogrooming and/or other contact) has previously been demonstrated to occur among ant nestmates [35] but not between host ants and their myrmecophiles. Even though previous studies have been founded on the assumption that myrmecophiles acquire rather than synthesise mimetic compounds to achieve chemical resemblance [36-38], acquisition has never been clearly demonstrated. A loss of host-specific surface compounds after isolation has already been demonstrated in the beetle *Myrmecaphodius excavaticollis* [36] as well as in the cricket *Myrmecophilus* sp. [39]. These results render the biosynthesis of host CHCs in these myrmecophiles unlikely, but a potential ability to down-regulate the biosynthesis of host-specific CHCs in the absence of a host cannot be ruled out. Such ability was found in the myrmecophilous butterfly *Phengaris (Maculinea) rebeli* [40].

Phengaris (Maculinea) rebeli caterpillars biosynthesise a subset of their host's hydrocarbons to become attractive, resulting in the transport into the nests [41]. Importantly, for this mechanism to work, the allomones produced by innate biosynthesis must be colony-unspecific. Indeed, *Phengaris (Maculinea)* caterpillars seem to mimic the surface chemistry of ant brood [41,42], which is generally less complex compared to that of workers and is assumed to be colony-unspecific. Hence, appropriate cues may mimic, for example, certain key stimuli of brood or males [38,40]. We presume that the more complex a host's recognition signature is, the more difficult it becomes for distantly related organisms to evolve the appropriate biosynthetic pathways for the production of the essential recognition cues and to express the compounds in the correct relative proportions (even if key regulatory enzymes are involved). In such cases, mixed strategies or the adoption of recognition cues may be evolutionarily more parsimonious. Another problem associated with the biosynthesis of mimetic cues is the dynamic nature of colony specific CHC profiles. An ant species is typically characterized by a set of CHCs, which differ among colonies in relative proportions [9]. Hydrocarbons are exchanged between nestmates by means of trophallaxis (exchange of nutritional liquids between nestmates) and allogrooming (grooming directed towards a nestmate), which establishes a uniform colony odour-the "gestalt odour" [9,43]. Despite its uniformity, the "gestalt odour" changes over time due to factors such as shifts in diet [44,45], different nesting materials [46] or seasonal differences [47]. Biosynthesis of worker CHCs is unlikely to be able to adjust to such flexible but specific "gestalt odours". These considerations may explain why an acquisition of mimetic CHCs is found more frequently than biosynthesis among distantly related parasites of social insects.

The role of accuracy in chemical mimicry

In addition to the mechanism of acquired chemical mimicry, our results highlight the importance of accuracy in chemical host resemblance by demonstrating that aggressive rejection can be avoided through closer chemical resemblance to the host. Notably, a parasites' successful social integration by chemical mimicry needs to include in principle only the cues that are necessary for nestmate recognition and not all the host CHCs. Nestmate recognition in the ant species *Formica exsecta*, for example, seems to be based only on selected compounds [48]. All types of compounds present on the cuticles of the host ant *L. distinguenda* and its parasitic silverfish could potentially be involved in recognition. Due to the generally accepted role of CHCs in ant nestmate recognition, we focused on non-polar compounds by using an appropriate solvent. There were several host CHCs on the silverfish, but only traces of other compounds (see additional file 2). Since host aggressiveness apparently depended on the chemical similarity of silverfish to their host, we conclude that the chemical recognition of silverfish by the host is predominantly based on CHCs. However, we were not able to differentiate which characteristics of CHC profiles, i.e. the composition (relative proportions), the presence or absence or the concentrations play the major role in the recognition of the silverfish.

A relationship between chemical resemblance and aggression is well known in the nestmate recognition of ants. Workers of the Argentine ant *Linepithema humile* showed elevated aggression against conspecific workers that were chemically more distant, while conspecific workers with similar profiles were treated amicably [49]. Among myrmecophiles, the *Phengaris (Maculinea) alcon* caterpillar biosynthesises a "pre-adoption" profile and adoption of caterpillars happened faster with higher accuracy of chemical mimicry of the host [42]. The innate biosynthesis of CHCs by myrmecophiles means that the origins of mimetic and model CHCs are different, which allows coevolutionary arms races to shape the degree of mimicry as well as the discrimination abilities by ants [50]. As described above, the synthesis of particular key stimuli used to deceive the host may be selected for in these scenarios [50]. These colony-unspecific stimuli allow the caterpillars to be adopted by any colony of their respective hosts, accompanied by local adaptation on a population level. In contrast, acquisition through physical contact to the host, as demonstrated here, means that the mimetic compounds of the model and the mimic are of identical origin. Coevolutionary arms races operate differently in this case, selecting for mimics with effective ways of acquiring host CHCs (e.g., through rubbing behaviour in *M. poneropila* or the consumption of host larvae in *Cosmophasis bitaeniata* [51]). In the host, selection favours defence mechanisms to prevent such "CHC pilfering" by parasitic myrmecophiles [52]. The present study indicates that such a coevolutionary arms race takes place between the host *L. distinguenda* and the myrmecophile *M. poneropila*. Sufficient contacts between the silverfish and the host ants are required to refresh the mimetic compounds and to gain increased chemical resemblance to the host, in order to acquire the colony's current "gestalt odour", resulting in social acceptance.

Besides adaptive adjustments in mimetic CHCs, the presence of additional cuticular compounds that do not match the ants' current template could potentially be responsible for recognition of aliens in social insect societies, and such cues could also explain the observed attacks against isolated silverfish. Workers of the carpenter ant *Camponotus herculeanus*, for example, attacked nestmates if they possessed one additional, foreign compound on their cuticle [45]. However, in our experiments an acquisition of additional compounds during isolation treatments that could have been responsible for the observed aggression seems unlikely for several reasons. First, aggression towards isolated and non-isolated host workers was not different, indicating that their chemical profiles were not influenced by the treatment. Second, we did not detect any specific compounds in silverfish (neither among isolated nor among non-isolated individuals) that could be responsible for the aggression, but we cannot exclude effects of compounds that were undetectable by the GC-MS analysis that was used. However, the silverfish isolation control experiment (additional file 6) finally shows that isolated silverfish did not acquire additional compounds during isolation that elicit aggression. Individuals that were first isolated and then were given the chance to re-acquire host CHCs were attacked significantly less than silverfish that were only isolated, indicating that only mimetic host compounds were behaviourally active.

Behavioural and morphological adaptations
Considering the level of integration a myrmecophile can achieve, we want to emphasise that mechanisms other than chemical integration may also play important roles, such as acoustic mimicry or behavioural and morphological adaptations [20,21,53]. The myrmecophilous cricket *Myrmecophilous formosanus*, for example, avoids ant attacks by swift movements [53]. *Malayatelura poneropila* was also regularly observed escaping by quick movements (behavioural category "avoid" in additional file 4). About 75% of the isolated silverfish survived the observation period despite frequent ant attacks during

escape. The limuloid (drop-shaped), scaled body of silverfish, with short appendages (antennae, cerci and praecerci) and retractable head, may also facilitate escaping ant attacks. The convergent evolution of the limuloid body form in unrelated myrmecophilous taxa provides strong evidence of its adaptive value [54]. These traits may also help *M. poneropila* to survive ant attacks in natural nests and perhaps offer the possibility of invading new host colonies, albeit this is presumably a risky manoeuvre. *Malayatelura poneropila* usually prefers central regions within natural nests where callows, pupae and larvae are located [26]. When individuals are able to reach this inner part of the nest, they are in a fairly safe place, which not only offers shelter and food, but also offers the possibility to steal the host's chemical profile by rubbing their surface on defenceless callows (see additional file 8: Video of *M. poneropila*).

Conclusion
In summary, we show that ant parasites can acquire CHCs directly from their host. Although elaborate behavioural adaptations may be required, the direct acquisition of host CHCs appears to be an evolutionarily parsimonious mechanism for taxonomically distant parasites such as *M. poneropila*. Furthermore, our study reveals that the accuracy of chemical mimicry can be crucial for parasites of social insects to gain social acceptance. For *M. poneropila*, regular replenishment of mimetic compounds increases survival because individuals with low chemical host resemblance are recognised and attacked frequently, sometimes captured and killed. Notably, the less frequently a silverfish replenishes its chemical profile (e.g., by failure to locate defenceless callows), the more difficult it becomes to remain unrecognised and to seek contact with the host ants.

Additional material

Additional file 1: Calculation of animal surface areas. Calculation of surface areas of the bodies of silverfish, workers and isopods.

Additional file 2: Table of compounds. Concentrations of 32 compounds that constituted 99.06% of the chemical profiles of workers ($N = 44$) across colonies evaluated by a similarity percentage analysis (SIMPER) on Bray-Curtis distances. In addition, concentrations of non-isolated (Sf 0 d; $N = 63$), six days isolated (Sf 6 d; $N = 12$) and nine day isolated silverfish (Sf 9 d; $N = 56$) across colonies are shown.

Additional file 3: Concentration of CHCs. Shown is the total quantity of surface chemicals per area of non-isolated and isolated silverfish.

Additional file 4: Behavioural interactions in the social acceptance experiment. Detailed information on behavioural interactions between silverfish and host ants across all colonies.

Additional file 5: NMDS plot of behavioural interactions between isolated and non-isolated silverfish and their host ants for colony 7.

Additional file 6: Silverfish isolation control experiment. Control for isolation treatment. Chemical similarities of silverfish to host workers and aggression toward the same individuals in colony 8.

Additional file 7: Worker control experiment. Concentration of CHCs on non-isolated and isolated workers.

Additional file 8: Video of *M. ponerophila*. The video shows *M. ponerophila* together with host workers and brood in an artificial nest site. One silverfish individual (in the foreground) rubs its own body intensely on that of an ant worker, presumably to acquire host CHCs.

Acknowledgements
We thank the behavioural ecology group at the LMU Munich for helpful comments on the manuscript, with special thanks to Sebastian Pohl. Many thanks go to Sofia Lizon à l'Allemand, Max Kölbl, Magdalena Mair, Hannah Kriesell and Deborah Schweinfest for their valuable assistance in the field. We are grateful for financial support from the DFG (Deutsche Forschungsgemeinschaft, project WI 2646/3) and for helpful comments of three anonymous referees.

Author details
[1]Department of Biology II, Ludwig Maximilian University Munich, Großhaderner Straße 2, Planegg-Martinsried, 82152, Germany. [2]Department of Organic Chemistry, Technical University Braunschweig, Hagenring 30, Braunschweig, 38106, Germany. [3]Institute of Biological Science, University Malaya, Kuala Lumpur, 50603, Malaysia.

Authors' contributions
CvB and VW designed the study, acquired the data, analysed and interpreted the data and drafted the manuscript. SS identified the chemical compounds and revised the manuscript. R H revised the manuscript. All authors approved the final manuscript.

Received: 26 August 2011 Accepted: 1 December 2011
Published: 1 December 2011

References
1. Dawkins R, Krebs JR: **Arms races between and within species.** *Proc R Soc B* 1979, **205**:489-511.
2. Combes C: **Arms races.** In *The art of being a parasite.* Edited by: Combes C. Chicago: The University of Chicago Press; 2005:8-20.
3. Combes C: *The art of being a parasite* Chicago: The University of Chicago Press; 2005.
4. Moore J: *Parasites and the behavior of animals* New York: Oxford University Press; 2002.
5. Wilson EO: *Success and dominance in ecosystems: the case of the social insects* Excellence in ecology, Oldendorf/Luhe: Ecology Institute; 1990.
6. Hölldobler B, Wilson EO: *The ants* Cambridge: Harvard University Press; 1990.
7. Vander Meer RK, Morel L: **Nestmate recognition in ants.** In *Pheromone communication in social insects: Ants, Wasps, Bees and Termites.* Edited by: Vander Meer RK, Breed M, Winston M, Espelie KE. Boulder:Westview Press; 1998:79-103.
8. Hefetz A: **The evolution of hydrocarbon pheromone parsimony in ants (Hymenoptera:Formicidae)-interplay of colony odor uniformity and odor idiosyncrasy. A review.** *Myrmecol News* 2007, **10**:59-68.
9. van Zweden JS, d'Ettorre P: **Nestmate recognition in social insects and the role of hydrocarbons.** In *Insect hydrocarbons: Biology, Biochemistry and Chemical Ecology.* Edited by: Blomquist GJ, Bagnères A.-G. New York: Cambridge University Press; 2010:222-243.
10. Lenoir A, D'Ettorre P, Errard C, Hefetz A: **Chemical ecology and social parasitism in ants.** *Annu Rev Entomol* 2001, **46**:573-99.
11. Blomquist GJ, Bagnères A.-G: *Insect Hydrocarbons: Biology, Biochemistry and Chemical Ecology* New York: Cambridge University Press; 2010.
12. Thomas JA, Schönrogge K, Elmes GW: **Specialisation and host association of social parasites of ants.** In *Insect Evolutionary Ecology.* Edited by: Fellowes M, Holloway G, Rolff J. London: CABI Publishing; 2005:475-514.
13. Schmid-Hempel P: *Parasites in social insects* Princeton: Princeton University Press; 1998.
14. Boomsma JJ, Schmid-Hempel P, Hughes WOH: **Life histories and parasite pressure across the major groups of social insects.** In *Insect Evolutionary Ecology.* Edited by: Fellowes MDE, Holloway GJ, Rolff J. Wellingford: CABI Publishing; 2005:139-175.
15. Wilson EO: *The insect societies* Cambridge: Harvard University Press; 1971.
16. Gösswald K: *Gäste der Ameisen.* In *Unsere Ameisen II.* Edited by: Gösswald K. Stuttgart: KOSMOS; 1955:.
17. Akre RD, Rettenmeyer CW: **Behavior of Staphylinidae associated with army ants (Formicidae:Ecitoninae).** *J Kans Entomol Soc* 1966, **39**(4):745-782.
18. Kistner DH: **Social and evolutionary significance of social insect symbionts.** In *Social insects.* Edited by: Hermann HR. New York: Academic press; 1979:339-413.
19. Dettner K, Liepert C: **Chemical mimicry and camouflage.** *Annu Rev Entomol* 1994, **39**:129-54.
20. Barbero F, Thomas JA, Bonelli S, Balletto E, Schönrogge K: **Queen ants make distinctive sounds that are mimicked by a butterfly social parasite.** *Science* 2009, **323(5915)**:782-785.
21. Dinter K, Paarmann W, Peschke K, Arndt E: **Ecological, behavioural and chemical adaptations to ant predation in species of *Thermophilum* and *Graphipterus* (Coleoptera: Carabidae) in the Sahara desert.** *J Arid Environ* 2002, **50(2)**:267-286.
22. Bagnères A-G, Lorenzi M: **Chemical deception/mimicry using cuticular hydrocarbons.** In *Insect hydrocarbons: Biology, Biochemistry and Chemical Ecology.* Edited by: Blomquist GJ, Bagnères A-G. New York: Cambridge University Press; 2010:282-324.
23. Akino T: **Chemical strategies to deal with ants: a review of mimicry, camouflage, propaganda, and phytomimesis by ants (Hymenoptera: Formicidae) and other arthropods.** *Myrmecol News* 2008, **11**:173-181.
24. Howard RW, Akre RD, Garnett WB: **Chemical mimicry in an obligate predator of carpenter ants (Hymenoptera: Formicidae).** *Ann Entomol Soc Am* 1990, **83**:607-616.
25. Vane-Wright RI: **On the definition of mimicry.** *Biol J Linn Soc* 1980, **13**:1-6.
26. Witte V, Foitzik S, Hashim R, Maschwitz U, Schulz S: **Fine tuning of social integration by two myrmecophiles of the ponerine army ant *Leptogenys distinguenda*.** *J Chem Ecol* 2009, **35(3)**:355-367.
27. Mendes L, von Beeren C, Witte V: ***Malayatelura ponerophila* - a new genus and species of silverfish (Zygentoma, Insecta) from Malaysia, living in *Leptogenys* army-ant colonies (Formicidae).** *Deut Entomol Z* .
28. Witte V, Leingärtner A, Sabaß L, Hashim R, Foitzik S: **Symbiont microcosm in an ant society and the diversity of interspecific interactions.** *Anim Behav* 2008, **76**:1477-1486.
29. Steghaus-Kovac S: **Wanderjäger im Regenwald-Lebensstrategien im Vergleich: Ökologie und Verhalten südostasiatischer Ameisenarten der Gattung *Leptogenys* (Hymenoptera: Formicidae: Ponerinae).** PhD thesis Frankfurt am Main: Johann Wolfgang Goethe Universität; 1994.
30. Jaisson P: **Social behavior.** In *Comprehensive insect physiology, biochemistry and pharmacology.* Edited by: Kerkut GA, Gilbert LI. Oxford: Pergamon Press; 1985:673-694.
31. Stein SE: **An integrated method for spectrum extraction and compound identification from GC/MS data.** *J Am Soc Mass Spectrom* 1999, **10**:770-781.
32. Witte V: **Organisation und Steuerung des Treiberameisenverhaltens bei südostasiatischen Ponerinen der Gattung *Leptogenys*** Dissertation, Johann Wolfgang Goethe University, Frankfurt am Main; 2001 [http//publikationen.ub.uni-frankfurt.de/files/5541/00000172.pdf].
33. Tanner CJ, Adler FR: **To fight or not to fight: context-dependent interspecific aggression in competing ants.** *Anim Behav* 2009, **77**:297-305.
34. Anderson MJ, Gorley RN, Clarke KR: *PERMANOVA+ for PRIMER: Guide to software and statistical methods* Plymouth: PRIMER-E; 2008.
35. Lenoir A, Hefetz A, Simon T, Soroker V: **Comparative dynamics of gestalt odour formation in two ant species *Camponotus fellah* and *Aphaenogaster senilis* (Hymenoptera: Formicidae).** *Physiol Entomol* 2001, **26**:275-83.
36. Vander Meer RK, Wojcik DP: **Chemical mimicry in the myrmecophilous beetle *Myrmecaphodius excavaticollis*.** *Science* 1982, **218**:806-808.
37. Vander Meer RK, Jouvenaz DP, Wojcik DP: **Chemical mimicry in a parasitoid (Hymenoptera: Eucharitidae) of fire ants (Hymenoptera: Formicidae).** *J Chem Ecol* 1989, **15**:2247-61.
38. Hojo MK, Wada-Katsumata A, Akino T, Yamaguchi S, Ozaki M, Yamaoka R: **Chemical disguise as particular caste of host ants in the ant inquiline parasite *Niphanda fusca* (Lepidoptera: Lycaenidae).** *Proc R Soc B* 2009, **276(1656)**:551-558.
39. Akino T, Mochizuki R, Morimoto M, Yamaoka R: **Chemical camouflage of myrmecophilous cricket *Myrmecophilus* sp. to be integrated with several ant species.** *Jpn J Appl Entomol Zool* 1996, **40**:39-46.

40. Schönrogge K, Wardlaw JC, Peters AJ, Everett S, Thomas JA, Elmes GW: Changes in chemical signature and host specificity from larval retrieval to full social integration in the myrmecophilous butterfly *Maculinea rebeli*. *J Chem Ecol* 2004, **30**:91-107.
41. Akino T, Knapp JJ, Thomas JA, Elmes GW: **Chemical mimicry and host specificity in the butterfly *Maculinea rebeli*, a social parasite of *Myrmica* ant colonies.** *Proc R Soc B* 1999, **266**:1419-1426.
42. Nash DR, Als TD, Maile R, Jones GR, Boomsma JJ: **A mosaic of chemical coevolution in a large blue butterfly.** *Science* 2008, **319(5859)**:88-90.
43. Crozier RH, Dix MW: **Analysis of two genetic models for the innate components of colony odor in social Hymenoptera.** *Behav Ecol Sociobiol* 1979, **4**:217-224.
44. Liang D, Silverman J: **"You are what you eat": Diet modifies cuticular hydrocarbons and nestmate recognition in the Argentine ant, *Linepithema humile*.** *Naturwissenschaften* 2000, **87**:412-416.
45. Guerrieri FJ, Nehring V, Jørgensen CG, Nielsen J, Galizia CG, d'Ettorre P: **Ants recognize foes and not friends.** *Proc R Soc B* 2009, **276**:2461-2468.
46. Heinze J, Foitzik S, Hippert A, Hölldobler B: **Apparent dear-enemy phenomenon and environment-based recognition cues in the ant *Leptothorax nylanderi*.** *Ethology* 1996, **102**:510-22.
47. Vander Meer RK, Saliwanchik D, Lavine B: **Temporal changes in colony cuticular hydrocarbon patterns of *Solenopsis invicta*: Implications for nestmate recognition.** *J Chem Ecol* 1989, **15(7)**:2115-2126.
48. Martin SJ, Vitikainen E, Helanterä H, Drijfhout F: **Chemical basis of nest-mate discrimination in the ant *Formica exsecta*.** *Proc R Soc B* 2008, **275**:1271-1278.
49. van Wilgenburg E, Torres CW, Tsutsui ND: **The global expansion of a single ant supercolony.** *Evol Appl* 2010, **3(2)**:136-143.
50. Nash A, Boomsma J: **Communiction between hosts and social parasites.** In *Sociobiology of communication an interdisciplinary approach.* Edited by: d'Ettorre P, Hughes DP. New York: Oxford University Press; 2008:55-79.
51. Elgar MA, Allan RA: **Predatory spider mimics acquire colony-specific cuticular hydrocarbons from their ant model prey.** *Naturwissenschaften* 2004, **91**:143-147.
52. von Beeren C, Maruyama M, Hashim R, Witte V: **Differential host defense against multiple parasites in ants.** *Evol Ecol* 2011, **25**:259-276.
53. Komatsu T, Maruyama M, Itino T: **Behavioral differences between two ant cricket species in Nansei Islands: host-specialist versus host-generalist.** *Insect Soc* 2009, **56(4)**:389-396.
54. Gotwald WH Jr: *Army ants: the biology of social predation* Ithaca: Cornell University Press; 1995.

doi:10.1186/1472-6785-11-30
Cite this article as: von Beeren *et al*.: **Acquisition of chemical recognition cues facilitates integration into ant societies.** *BMC Ecology* 2011 **11**:30.

Submit your next manuscript to BioMed Central and take full advantage of:

• Convenient online submission
• Thorough peer review
• No space constraints or color figure charges
• Immediate publication on acceptance
• Inclusion in PubMed, CAS, Scopus and Google Scholar
• Research which is freely available for redistribution

Submit your manuscript at
www.biomedcentral.com/submit

Supplementary Material

Title: Acquisition of chemical recognition cues facilitates integration into ant societies

Christoph von Beeren, Stefan Schulz, Rosli Hashim, & Volker Witte

Additional file 1 – Calculation of animal surface areas.

Silverfish. To estimate the surface area of a silverfish, we approximated their body form by dividing the body into different geometrical parts. The head area was estimated by calculating a quarter of the surface area of a sphere (dorsal surface) plus half the surface area of a circle (flat, ventral surface). The rest of the body was estimated by calculating half of the surface area of a cone (dorsal surface) plus the area of an isosceles triangle (flat, ventral surface). The picture of *M. ponerophila* shows that the simplified body form approximately matches the actual body form. Accordingly, the surface area of each silverfish was calculated using the following formula (Dorn et al. 2005):

$$\text{Surface area}_{\text{silverfish}} = \frac{4\pi r^2}{4} + \frac{\pi r^2}{2} + \frac{\pi r s}{2} + \frac{2hr}{2}$$

Overall, we measured the surface area of 180 individuals (including the 90 individuals used for calculating the surface concentration of CHCs). The median surface area was 13.48 mm^2 with a maximum of 19.71 mm^2 and a minimum of 2.44 mm^2. The data set shows no normal distribution (Shapiro-Wilk, $W = 0.98$, $P < 0.005$).

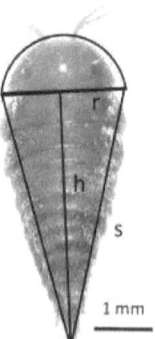

Ant workers. The worker's body shape was divided into geometrical parts to estimate their surface area. The surface area of the head, the allitrunk (thorax), the petiole and the gaster were calculated by using the approximation of a three-axis ellipsoid according to Thomson (http://de.wikipedia.org/wiki/Ellipsoid):

$$\text{Surface area}_{\text{ellipsoid}} = 4\pi \left[\frac{(ab)^{1.6} + (ac)^{1.6} (bc)^{1.6}}{3} \right]^{0.625}$$

The variables a, b and c correspond to the length, the breadth and the height of the respective ant body part. The lateral surface area of four circular cylinders was calculated to estimate the surface area of the legs consisting of the coxa, the femur, the tibia and the tarsus (the surface area of the trochanter was neglected).

$$\text{Surface area}_{\text{lateral area of cylinder}} = 2\pi r h = \pi d h$$

The variables d and h correspond to the length and breadth of the respective part of the ants' leg. We measured the surface area of 10 individuals. The median surface area was 78.24 mm^2 with a maximum of 83.09 mm^2 and a minimum of 71.76 mm^2. The data set shows a normal distribution (Shapiro-Wilk, $W = 0.92$, $P = 0.400$).

Isopods. The area of an isopod's dorsal surface was calculated by halving the surface area of a three-axis ellipsoid according to Thomson (see above). The area of the isopod's ventral surface was determined by applying the formula of an ellipse:

$$\text{Surface area}_{\text{ellipse}} = 2\pi ab$$

The variables a and b represent half of the ellipse's major and minor axes, respectively. The median surface area of isopods ($N = 22$) was 42.69 mm^2 with a maximum of 95.86 mm^2 and a minimum of 27.04 mm^2. The data set shows no normal distribution (Shapiro-Wilk, $W = 0.83$, $P = 0.017$).

Reference:
Dorn H-J, Freudigmann H, and Herbst M (2005). **Formelsammlung Mathematik. Gymnasium: Sekundarstufe I und II.** Ernst Klett Verlag, Stuttgart, GER

Additional file 2 – Table of compounds. Concentrations of 32 compounds that constituted 99.06% of the chemical profiles of workers ($N = 44$) across colonies evaluated by a similarity percentage analysis (SIMPER) on Bray-Curtis distances. In addition, concentrations of non-isolated (Sf 0 d; $N = 63$), six days isolated (Sf 6 d; $N = 12$) and nine days isolated silverfish (Sf 9 d; $N = 56$) across colonies are shown. Abbreviation: SE = standard error

Compound	Workers	Concentrations [µg/mm² ± SE]		
		Sf 0 d	Sf 6 d	Sf 9 d
Nonacosene (A)	18.685 ± 1.119	7.254 ± 0.926	3.222 ± 0.782	4.715 ± 0.780
Tricosane	17.830 ± 0.862	14.229 ± 1.000	4.441 ± 0.895	5.043 ± 0.655
Heptacosene (A)	13.618 ± 1.039	5.318 ± 0.702	1.373 ± 0.400	3.558 ± 0.618
Pentacosene (A)	11.812 ± 1.025	3.522 ± 0.367	0.536 ± 0.127	1.808 ± 0.294
Pentacosadiene	8.280 ± 0.489	3.159 ± 0.551	0.843 ± 0.187	1.583 ± 0.342
Hentriacontene	6.016 ± 0.620	0.502 ± 0.084	0.076 ± 0.045	0.078 ± 0.026
Pentacosene (B)	5.686 ± 0.381	2.321 ± 0.305	0.770 ± 0.167	1.289 ± 0.218
Heptacosadien	5.155 ± 0.496	1.135 ± 0.269	0.092 ± 0.062	0.482 ± 0.203
Heptacosene (B)	3.683 ± 0.554	0.810 ± 0.113	0.207 ± 0.070	0.485 ± 0.090
Pentacosane	3.512 ± 0.192	2.658 ± 0.235	0.923 ± 0.191	0.977 ± 0.142
Nonacosene (B)	1.958 ± 0.206	0.170 ± 0.039	0.085 ± 0.052	0.028 ± 0.017
Pentacosene (C)	1.383 ± 0.111	0.774 ± 0.144	0.163 ± 0.084	0.316 ± 0.077
Tricosene (C)	1.072 ± 0.175	0.826 ± 0.162	0.118 ± 0.061	0.230 ± 0.059
11-Methylpentacosane	1.036 ± 0.091	0.206 ± 0.034	0.113 ± 0.055	0.052 ± 0.017
Decyloctanoate	0.904 ± 0.100	0.004 ± 0.003	-	0.007 ± 0.007
Tritriacontene	0.847 ± 0.125	0.001 ± 0.001	-	-
Tricosene (B)	0.833 ± 0.074	0.093 ± 0.023	0.008 ± 0.008	0.002 ± 0.002
Nonacosadiene	0.832 ± 0.159	-	0.025 ± 0.025	-
11-Methylheptacosane	0.806 ± 0.063	0.239 ± 0.035	0.309 ± 0.131	0.062 ± 0.018
Tetracosane	0.651 ± 0.037	0.503 ± 0.074	0.156 ± 0.047	0.149 ± 0.032

Chapter 2: Supplementary material

Compound				
Decyl decanoate	0.610 ± 0.072	0.001 ± 0.001	-	0.002 ± 0.011
Octacosene (A)	0.505 ± 0.064	-	-	-
Multiple methylated hentriacontenes	0.494 ± 0.102	-	-	-
13- and 15- Methylhentriacontanes	0.450 ± 0.087	-	-	-
6,9,12,15-Heptacosatetraene	0.432 ± 0.049	0.033 ± 0.012	-	0.005 ± 0.003
13- and 15- Methylnonacosanes	0.394 ± 0.056	0.001 ± 0.001	0.056 ± 0.052	-
Docosane	0.307 ± 0.018	0.449 ± 0.136	0.042 ± 0.013	0.005 ± 0.005
Octacosene (B)	0.279 ± 0.058	0.003 ± 0.003	-	-
Hexacosene (B)	0.254 ± 0.057	0.001 ± 0.000	-	-
Heneicosane	0.196 ± 0.021	0.280 ± 0.071	0.023 ± 0.008	0.001 ± 0.000
9-Methyltricosane	0.132 ± 0.010	0.001 ± 0.001	-	-

Additional file 3 – Concentration of CHCs. The graph shows the total quantity of surface chemicals per area of non-isolated and isolated silverfish. One outlier of a non-isolated silverfish in colony 1 is not shown in the graph for better visualisation (outlier = 171 ng/mm2). Significant differences between groups were evaluated by PERMANOVA (***$P < 0.001$; **$P < 0.010$; n.s. = not significant). Median (+= mean), quartiles (boxes), 10th and 90th percentiles (whiskers), and outliers (♦ = outlier) are shown. Abbreviations: No iso = no isolation, d iso = days isolated

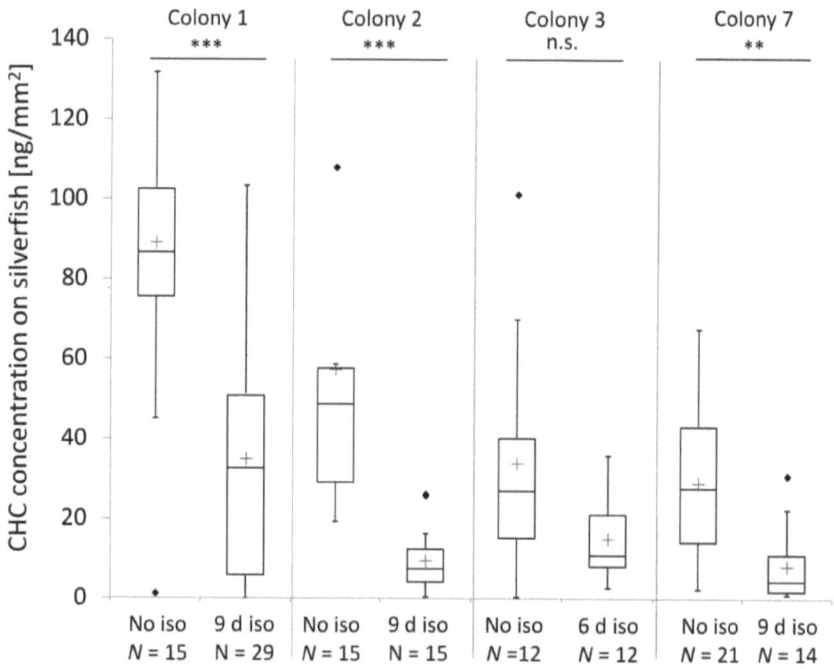

Additional file 4 – Behavioural interactions in the social acceptance experiment.

Interaction	Colony 4 No isolation $N=7$		Colony 4 6d isolation $N=6$		Colony 5 No isolation $N=18$		Colony 5 6d isolation $N=22$		Colony 7 No isolation $N=21$		Colony 7 9d isolation $N=12$	
Ignored	59 / 9	A	50 / 8.5	A	219 / 12.5	A	101 / 4	B	267 / 10	A	82 / 6	B
Groomed	6 / 0	A	0 / 0	A	8 / 0	A	1 / 0	A	0 / 0	A	1 / 0	A
Avoid	4 / 0	A	28 / 5.5	B	35 / 0	A	170 / 2	B	98 / 4	A	149 / 12.5	B
Antennated	3 / 0	A	15 / 2	B	17 / 0	A	45 / 1	B	34 / 1	A	33 / 3	B
Unnoticed	260 / 40	A	156 / 25	B	623 / 35	A	326 / 14	B	659 / 31	A	257 / 21	B
Chased	1 / 0	A	30 / 4.5	B	1 / 0	A	185 / 8	B	1 / 0	A	34 / 3	B
Snapped	0 / 0	A	23 / 3.5	B	6 / 0	A	291 / 11.5	B	25 / 1	A	39 / 3	B
Stung	0 / 0	A	6 / 1	B	1 / 0	A	24 / 1	B	0 / 0	A	1 / 0	A

The upper number in each array represents the sum and the lower number indicates the median of the corresponding interaction. Different capital letters depict significant differences ($P < 0.05$) for a given behavioural interaction evaluated by PERMANOVA. We did not apply statistics for colony 6 because none of the isolated individuals completed the standardized number of 50 ant contacts (Tab. 1). Abbreviations: N = number of silverfish, 6d = six days, 9d = nine days

Additional file 5 – NMDS plot of behavioural interactions between isolated and non-isolated silverfish and their host ants for colony 7. Each data point represents 50 encounters of a silverfish individual with a host worker. Arrows represent the relative contributions of behaviours (see Table 2) to data separation, whereby the length indicates the importance (observed frequency). For clarity, the origin of arrows is not centred in the plot. "Stress" is a quality measure of NMDS.

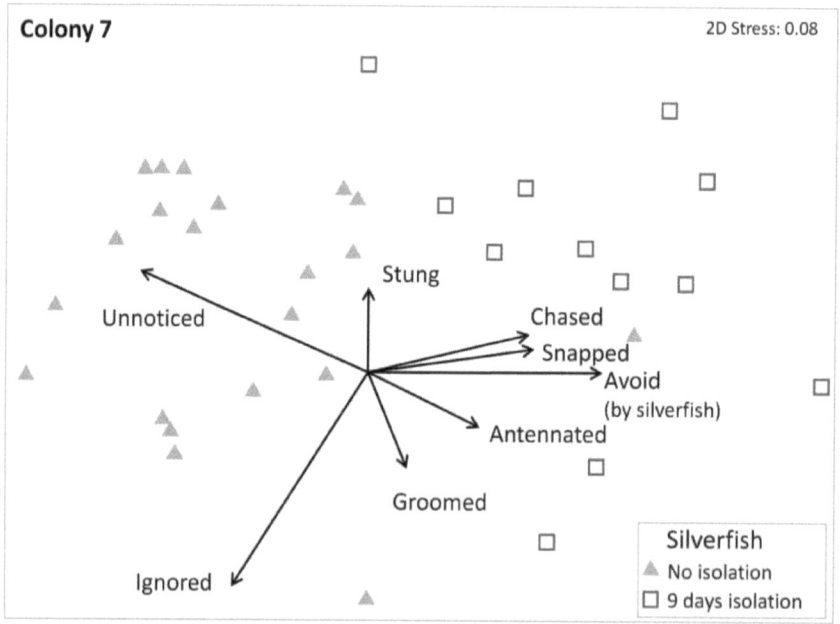

Additional file 6 − Silverfish isolation control experiment. Chemical similarities of silverfish to host workers and aggression toward the same individuals in colony 8. Differences between groups were evaluated by PERMANOVA (***$P < 0.001$). Median (+ = mean), quartiles (boxes), 10th and 90th percentiles (whiskers), and outliers (♦ = outlier) are shown. Abbreviations: 6 d iso = 6 days isolation, 1 d Wo = silverfish kept one day together with host workers

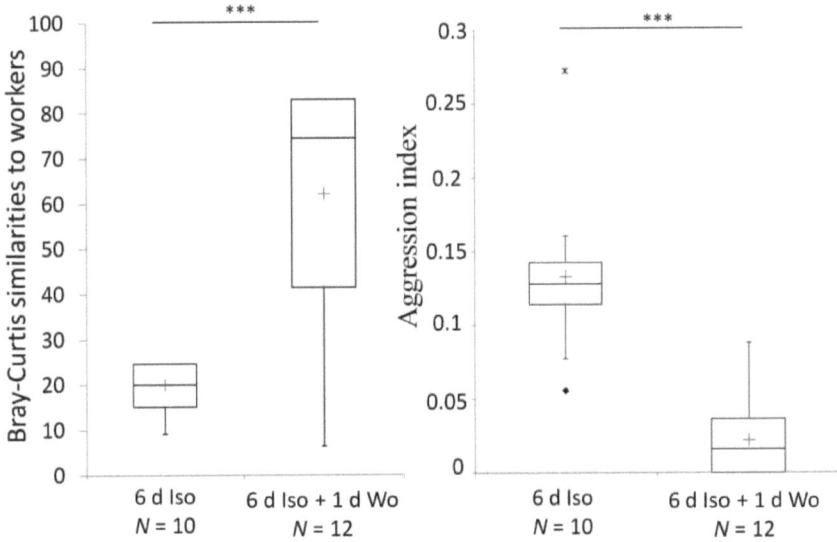

Additional file 7 – Worker control experiment. Concentration of CHCs on non-isolated and isolated workers. Significant differences between groups were evaluated by PERMANOVA (*$P < 0.05$; ***$P < 0.001$; n.s. = not significant). Median (+ = mean), quartiles (boxes), 10th and 90th percentiles (whiskers), and outliers (♦ = outlier) are shown. Abbreviations: 9 d iso = nine days isolation

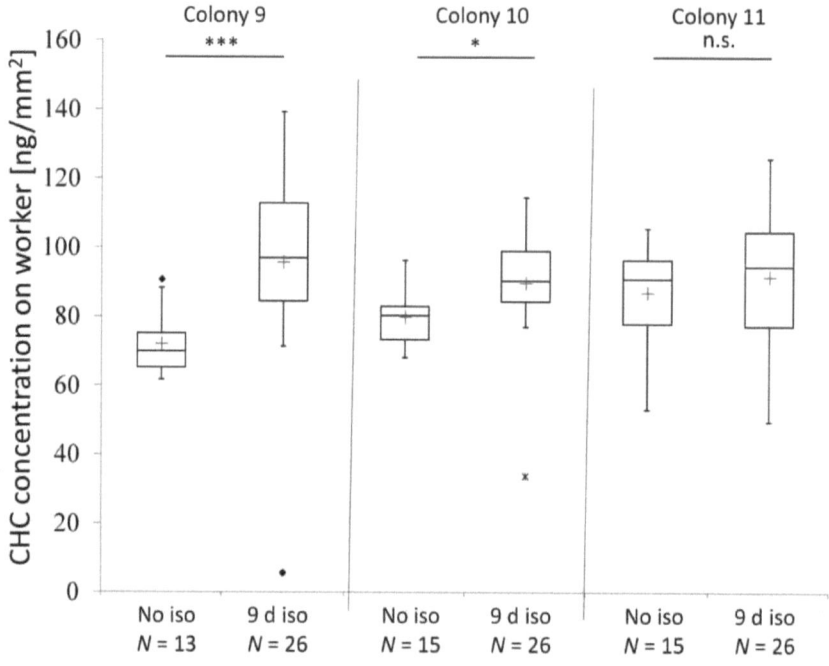

Video. A video is available on the following website: http://www.biomedcentral.com/1472-6785/11/30/additional

The video shows *M. ponerophila* together with host workers and brood in an artificial nest site. One silverfish individual (in the foreground) rubs its own body intensely on that of an ant worker, presumably to acquire host CHCs.

Chapter 3

The social integration of a myrmecophilous spider does not depend exclusively on chemical mimicry

Christoph von Beeren, Rosli Hashim and Volker Witte

Gamasomorpha maschwitzi *individuals are often found on top of host workers and rub their legs intensively on host bodies to steal the hosts' scent.* © *Christoph von Beeren*

2012

♦

von Beeren C, Schulz S, Hashim R and Witte V (2012). The social integration of a myrmecophilous spider does not depend exclusively on chemical mimicry. *The Journal of Chemical Ecology,* 38:262 – 271, DOI 10.1007/s10886-012-0083-0

J Chem Ecol (2012) 38:262–271
DOI 10.1007/s10886-012-0083-0

The Social Integration of a Myrmecophilous Spider Does Not Depend Exclusively on Chemical Mimicry

Christoph von Beeren · Rosli Hashim · Volker Witte

Received: 2 November 2011 / Revised: 15 February 2012 / Accepted: 16 February 2012 / Published online: 18 March 2012
© Springer Science+Business Media, LLC 2012

Abstract Numerous animals have evolved effective mechanisms to integrate into and exploit ant societies. Chemical integration strategies are particularly widespread among ant symbionts (myrmecophiles), probably because social insect nestmate recognition is predominantly mediated by cuticular hydrocarbons (CHCs). The importance of an accurate chemical mimicry of host CHCs for social acceptance recently has been demonstrated in a myrmecophilous silverfish. In the present study, we investigated the role of chemical mimicry in the myrmecophilous spider *Gamasomorpha maschwitzi* that co-occurs in the same host, *Leptogenys distinguenda*, as the silverfish. To test whether spiders acquire mimetic CHCs from their host or not, we transferred a stable isotope-labeled hydrocarbon to the cuticle of workers and analyzed the adoption of this label by the spiders. We also isolated spiders from hosts in order to study whether this affects: 1) their chemical host resemblance, and 2) their social integration. If spiders acquired host CHCs, rather than biosynthesizing them, they would be expected to lose these compounds during isolation. Spiders acquired the labeled CHC from their host, suggesting that they also acquire mimetic CHCs, most likely through physical contact. Furthermore, isolated spiders lost considerable quantities of their CHCs, indicating that they do not biosynthesize them. However, spiders remained socially well integrated despite significantly reduced chemical host similarity. We conclude that *G. maschwitzi* depends less on chemical mimicry to avoid recognition and aggressive rejection than the silverfish previously studied, suggesting that the two myrmecophiles possess different adaptations to achieve social integration.

Keywords Acquired chemical mimicry · Myrmecophile · Social integration · Cuticular hydrocarbons · *Malayatelura ponerophila*

Introduction

The phenomenon of mimicry was first described by the English naturalist Henry Walter Bates, who discovered that nontoxic species of Neotropical butterflies visually resemble toxic species, thus avoiding predation (Bates, 1862). Since this discovery, visual resemblance and its evolutionary consequences have been studied extensively (Müller, 1878; Fisher, 1927; Wickler, 1968; Brower, 1988; Ruxton et al., 2004). While scientists initially focused solely on visual mimicry, many researchers now also study chemical mimicry (Dettner and Liepert, 1994; Bagnères and Lorenzi, 2010).

Chemical communication is the most widespread form of communication among organisms (Symonds and Elgar, 2008; Steiger, et al., 2011), and social insects base their communication to a great extent on chemicals (Wilson, 1990). This includes a sophisticated chemical recognition system, able to distinguish group members from aliens, which helps protect societies from exploitation. In ants, wasps, bees, and termites, recognition of group members is based mainly on complex blends of cuticular hydrocarbons (CHCs) (van Zweden and

Electronic supplementary material The online version of this article (doi:10.1007/s10886-012-0083-0) contains supplementary material, which is available to authorized users.

C. von Beeren · V. Witte (✉)
Department of Biology II,
Ludwig-Maximilians University Munich,
Großhaderner Str. 2,
82152 Planegg-Martinsried, Germany
e-mail: witte@bio.lmu.de

R. Hashim
Department of Biological Sciences, University Malaya,
Kuala Lumpur, Malaysia

d'Ettorre, 2010). However, various arthropods prevent being recognized as aliens by mimicking the CHCs of social insect workers (chemical mimicry, sensu Dettner and Liepert, 1994) and appearing to be nestmates (Bagnères and Lorenzi, 2010). The origin of the CHCs used by mimics is unknown in the majority of cases; they may acquire CHCs directly from the host ('acquired chemical mimicry', sensu von Beeren et al., 2012), they may actively biosynthesize recognition cues ('innate chemical mimicry', sensu von Beeren et al., 2012), or both mechanisms may occur in combination (Akino, 2008).

Regardless of the origin of mimetic compounds, studies of similarities in CHC profiles between parasites and social insect hosts remain predominantly descriptive. A chemical resemblance alone does not necessarily mean that the host is deceived by the mimic, or that the mimic gains benefits through chemical resemblance. Specific bioassays are necessary to demonstrate whether chemical resemblance indeed affects the behavior of the host, as has been demonstrated in the spider *Cosmophasis bitaeniata* (Allan et al., 2002), the butterfly *Phengaris (Maculinea) alcon* (Nash et al., 2008) and the silverfish *Malayatelura ponerophila* (von Beeren et al., 2011b).

In this study, we investigated the role of chemical mimicry in the social integration of the kleptoparasitic spider, *Gamasomorpha maschwitzi* (Arachnida: Aranea: Oonipidae), which parasitizes the ponerine army ant *Leptogenys distinguenda*. Many spiders live in close relationship with ants, but little is known about how they specifically adapt to cope with host defenses (reviewed in Cushing, 1997). Species have been shown to be behaviorally adapted (Ceccarelli, 2007), chemically integrated (Allan et al., 2002), or to utilize both strategies together (Witte et al., 2009). Witte et al. (2009) suggested that, in the spider *G. maschwitzi*, additional factors to chemical mimicry might play roles. Although the spiders were chemically less similar to the host than was another myrmecophile, the silverfish *M. ponerophila*, they showed comparable integration and received fewer attacks. Apparently, integration of spiders depended less on chemical cues and more on other mechanisms.

The goals of the present study were twofold. We aimed to answer the question of whether *G. maschwitzi* acquires CHCs from the host or biosynthesizes them. Furthermore, we tested whether the accuracy of chemical host resemblance influences the level of social integration, as previously shown in the silverfish *M. ponerophila*. Silverfish individuals that were chemically more similar to host ants were attacked less often, and thus achieved a higher level of social integration (von Beeren et al., 2011b). We expected that: (1) spiders acquire CHCs from host ants through frequent contact (Witte et al., 2009); and (2) that spiders more closely resembling the host profile would be better integrated socially. To test these predictions, we performed two experiments. First, we applied a stable isotope-labeled hydrocarbon to the cuticle of ants and monitored the transmission of this label to the myrmecophilous spider. We expected that, if *G. maschwitzi* acquired CHCs from its host, the label would also accumulate on their cuticles. Second, we aimed to reduce the chemical resemblance of spiders to the host, so as to study the effects on social integration. For this, we isolated spiders from hosts for 9 days, assuming that spiders acquire host CHCs behaviorally (Witte et al., 2009), and therefore assumed that they would lose host compounds, resulting in reduced chemical similarity to the host. We hypothesized that spiders exhibiting reduced chemical host resemblance would be less socially integrated, as a consequence of being recognized as alien more often. We tested whether isolated spiders (with reduced host similarity) were attacked more frequently than unmanipulated individuals. Finally, we discuss here the integration strategies of the spider *G. maschwitzi* and the silverfish *M. ponerophila*.

Methods and Materials

Field Collections

Animals were collected and observed at the Field Studies Centre of the University of Malaya in Ulu Gombak, Malaysia (03°19.479' N, 101°45.163' E, altitude 230 m) and at the Institute of Biodiversity in Bukit Rengit, Malaysia (03° 35.779' N, 102°10.814' E, altitude 72 m). A total of 14 mo of fieldwork was carried out from August 2007–April 2011. The two kleptoparasites, *G. maschwitzi* (Wunderlich, 1994) and *M. ponerophila* (Mendes et al., 2011), both parasitize the nocturnal, ponerine army ant *L. distinguenda* (Emery, 1887). We searched for host nest sites during the night, between 8 p.m. and 3 a.m., by backtracking the ants' raiding trails. Subsequently, we checked nest sites every 30 min. for the onset of a colony migration. Host colonies frequently move to new nest sites (on average every 1.5 nights; Steghaus-Kovac, 1994). Under natural conditions, spiders participate in migrations by showing a "tandem-running"-like behavior (Witte et al., 1999; Fig. 1), while silverfish are phoretically transported on ant pupae, as well as running among ant workers (Witte et al., 2008). Using aspirators, we collected host ants during raids, and spiders and silverfish during migrations. Animals that were kept over several days in the laboratory (see below) were fed every day with freshly killed crickets. Crickets are among the natural diet of the host ants (Steghaus-Kovac 1994), the spiders and the silverfish (personal observation). Since *G. maschwitzi* occurs in low numbers, experimental procedures frequently were limited for working with appropriate sample sizes.

Chemical Transfer Experiment

The aims were twofold: first, we tested whether spiders can acquire CHCs from their host; second, we compared the quantity of the transferred label from host to spiders and

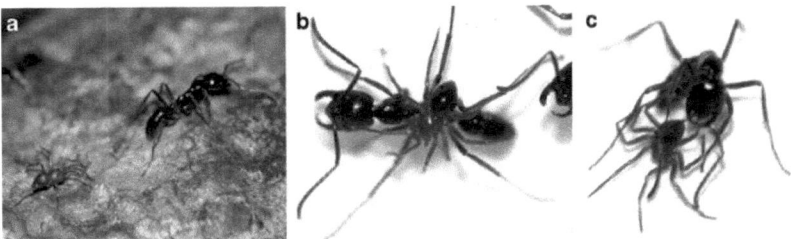

Fig. 1 (**a**) *Gamasomorpha maschwitzi* participates in ant migrations by performing a "tandem-running"–like behavior. The spiders are easily distinguishable from the host ants due to their reddish color. (**b**) In laboratory colonies, spiders frequently interact with ant workers and are often found on top of them. (**c**) Occasionally, a spider is recognized as an intruder and attacked by ant workers

silverfish. All animals were kept in an artificial nest (plastic box: 30×20×15 cm), constructed of soil and leaf litter from the environment; we used 65 ant workers, 65 callows, approx. 40 larvae, 30 pupae, 17 spiders, 12 silverfish, as well as 11 non-myrmecophilous isopods as controls. Workers and callows were treated with a stable isotope-labeled hydrocarbon (eicosane-D_{42}, C/D/N Isotopes Inc., Pointe-Claire Canada). We used eicosane-D_{42}, because of its similar properties (chain length, molecular weight) to the CHCs that occur naturally on the host (Table 1 of supplement). In a clean 20-ml glass vial, 200 μl of a saturated eicosane-D_{42} hexane solution were evaporated so that the hydrocarbon fully covered the bottom of the vial as a solid film. Workers and callows then were placed in the vial, and shaken gently for 30 min. to transfer the labeled compound. Ants did not noticeably suffer from this treatment. Labeled ants, spiders, silverfish, and control isopods were kept together for 3 d in the laboratory nest and, subsequently, 20 ants (10 workers and 10 callows) and all spiders, silverfish, and control isopods were extracted for chemical analysis. Non-myrmecophilous isopods, collected from the same rainforest, were added to test whether eicosane-D_{42} transmits to animals that are not in direct contact with the host. Preliminary studies revealed that isopods rarely had contact with ants and survived well in laboratory nests (von Beeren et al., 2011b). Therefore, they were well suited as controls. Importantly, their cuticle could adsorb the labeled CHC (von Beeren et al., 2011b).

Isolation Treatments

To evaluate the impact of isolation on CHC profiles, and on behavioral interactions with host ants, some spiders were separated from their home colony and kept isolated for 9 d in plastic boxes (21×15×5 cm), with a moistened plaster ground (for sample sizes refer to Table 1). The cuticles of isolated and non-isolated spiders were extracted with hexane to analyze differences in CHCs (colonies 1–4; see section "Comparison of CHCs"). In colony 4, we also tested the social acceptance by hosts of isolated and non-isolated spiders before extraction (see methods, section "Social Acceptance Experiment"). The chemical studies were performed independently (colonies 1–3) of, and in combination with (colony 4), behavioral observations, because the combined experiment is best suited to test whether an individual's chemical host resemblance affects its level of social acceptance. Nevertheless, an increase of CHC quantity through behavioral interactions with host ants (e.g., rubbing behavior) during the social acceptance experiment (see below) cannot be ruled out. Hence, we also extracted the cuticles of spiders that were not subjected to the social acceptance experiment.

Comparison of CHCs between Isolated and Non-isolated Spiders

To analyze whether isolation changed concentration, presence or absence, and/or composition of CHCs in spiders, we extracted the cuticles of individuals directly after collecting them in the field, and after 9 d of separation from the host (colony 1–4) (for sample sizes, see Table 1). Animals were transferred individually into 2-ml vials, with

Table 1 Overview of the number of spider individuals observed within each colony for the analyses of cuticular hydrocarbons (CHCs) and the social acceptance experiments

Colony	Analysis of CHCs		Social Accept. Exp.	
	Sp0d	Sp9d	Sp0d	Sp9d
Colony 1	9	6	-	-
Colony 2	11	5	-	-
Colony 3	9	4	-	-
Colony 4	13	10	13	10
Colony 5	-	-	9	-
Colony 6	-	-	5	-
Colony 7	-	-	5	-
Colony 8	-	-	10	-

Abbreviations: Sp0d=non-isolated spiders; Sp9d=9 d-isolated spiders

a polytetrafluoroethylene (PTFE) septum, and extracted for 10 min. in 200 μl hexane (HPLC grade, Sigma-Aldrich). The solvent was evaporated, and the CHCs were dissolved in 40 μl of hexane, containing an internal standard (methyl stearate, FLUKA Analytics, Sigma-Aldrich). Each sample (1 μl) was analyzed by gas chromatography–mass spectrometry (Agilent Technologies 6890 N-5975), using a Restek Rxi-5MS column (30 m length, 0.25 mm ID, 0.25 μm film thickness), at LMU Munich, Germany. Details of the methods can be found in Witte et al. (2009).

Chemicals were identified by their mass spectra and retention indices (RI), and peak areas calculated using the software AMDIS (version 2.68) (Stein, 1999). A target library of 109 compounds was created, based on compounds found in extracts of host ants and myrmecophiles (a list of identified compounds is given in Witte et al., 2009). Since AMDIS uses the mass spectrum as well as the retention index to identify a compound, it has the advantage of reliably detecting compounds, even at low quantities.

The absolute quantity per surface area of an individual (concentration) of each compound was calculated against the internal standard (20 ng/μl). We divided the total quantity of compounds by the median surface area, in mm^2, for workers, spiders, silverfish, and isopods, to standardize to the presumed concentration of surface compounds at the point of an ant's antennal contact. To determine surface area, a spider's body was divided into geometrical parts, and the relevant body dimensions were measured using a stereomicroscope (Zeiss Stemi 2000-C) with a measuring eyepiece (see supplemental material: Calculation of surface area). We used the calculated surface area of workers, spiders, and silverfish according to von Beeren et al. (2011b). The median surface areas of animals are: workers (median=78.24 mm^2, range=71.76–83.09 mm^2, N=10), spiders (median=38.19 mm^2, range=32.78–42.03 mm^2, N=15), silverfish (median=13.48 mm^2, range=2.44–19.71 mm^2, N=180), and isopods (median=42.69 mm^2, range=27.04–95.86 mm^2, N=22).

Social Acceptance Experiment

In order to evaluate the social acceptance of non-isolated and isolated spiders, we quantified the aggression of hosts against individual spiders, by performing a contact study in laboratory nests (for sample sizes, see Table 1). The behavioral responses of host workers to spiders were studied under laboratory conditions, using artificial nests, consisting of a transparent plastic container (20×14×1 cm) with a 1.5 cm nest entrance shaded with a plastic cover. The artificial nests were placed in a larger foraging arena (32×25×9 cm), with a moistened plaster ground to maintain humidity. Laboratory nests contained 200 mature host ant workers collected in raids, as workers from raids are more likely to defend the colony compared to young callows that

stayed in the nest (CvB, personal observation). Before introducing spiders, host colonies were given 1 h to settle in the artificial nest, because ants tend to be more aggressive in familiar than in unfamiliar areas (Tanner and Adler, 2009). Spiders were tested either within 6 h of collection from the field, or after 9 d of isolation. The interactions of ant workers with one focal spider was observed in approximately 50 consecutive encounters, by recording eight different behaviors (Table 2). Each spider was tested only once. Repeated interactions with the same ants were possible. However, since we were interested in defense at the colony-level against spiders, and since task allocation occurs naturally in ants, repeated interactions do not affect our interpretations.

An aggression index (AI) was calculated for each spider from the observed interactions. The AI focused on the proportion of the aggressive ant reactions, chased, snapped and stung: $AI=N_A/N_T$; with N_A=number of aggressive interactions and N_T=total number of interactions.

Combined Experiment

To test whether a relationship between chemical similarity and ant aggression exists, we studied both social acceptance and analysis of CHCs for one *L. distinguenda* colony (colony 4; Table 1). Host aggression was first quantified for each spider, via the social acceptance experiment, and then

Table 2 Behavioral interactions between spiders and ants, and categories used for calculating the aggression index

Behavior	Definition	Category
Ignored	An ant worker touches the spider once with its antennae and moves on without any sign of behavioral modification.	-
Groomed	An ant grooms the spider with its mouthparts. The spider remains in position.	-
Avoid	When an ant approaches, the spider avoids contact by quick escape.	-
Antennated	An ant touches a spider repeatedly with its antennae for longer than 2 sec without displaying other behaviors.	-
Unnoticed	An ant comes into, and perhaps stays in contact with a spider, but not with its antennae; the ant does not modify its behavior.	-
Chased	An ant touches the spider with its antennae and quickly lunges in its direction.	Aggressive
Snapped	An ant touches the spider with its antennae and snaps with its mandibles into the direction of the spider.	Aggressive
Stung	An ant touches the spider with its antennae, lunges forward and bends its gaster in direction of the opponent. The attempt does not need to be successful.	Aggressive

cuticular chemicals were extracted. We performed this experiment with non-isolated and nine-days isolated spiders.

Comparison between the Spider and the Silverfish

The experiments were identical to those of a study on the myrmecophilous silverfish *M. ponerophila* (von Beeren et al., 2011b). Consequently, we compared the results of the study on the spider with those on the silverfish (von Beeren et al., 2011b).

Data Analysis

Chemical and behavioral data were evaluated using PRIMER 6 (version 6.1.12, Primer-E Ltd., Ivybridge, U.K.) with the PERMANOVA + add-in (version 1.0.2) (Anderson et al., 2008), using a non-parametric permutational analysis of variance (PERMANOVA) with 9,999 permutations. PERMANOVA Analyses were based on Bray-Curtis similarities (as a semi-metric measure) or simple matching (a presence-absence measure), either calculated from a single response variable (chemical similarity, CHC concentration, aggression index), or from numerous response variables (CHC profiles, presence or absence of CHCs, behavioral interactions). Box plots were created from univariate data with the Microsoft Excel add-in SSC-Stat (version 2.18, Statistical service centre of the University of Reading, Reading, U.K.).

Chemical Analysis

As no spider-specific compounds were detected, only the principal compounds that contributed, in total, to 99% of the chemical profiles of workers (N=49), according to a similarity percentage analysis (SIMPER) on Bray-Curtis similarities, were used in the statistical analysis. These selected data (N=32 compounds; supplement Table 1) were used to analyze the CHC composition of spiders, the presence or absence of CHCs, the total CHC concentration, and the chemical similarity of spiders to workers. To test whether the chemical similarity of spiders to host colony was influenced by isolation, Bray-Curtis similarities to the average worker CHC profile of the respective host colony were used as a univariate response variable, and a PERMANOVA with a 2-factor nested design [colonies (random), days of isolation (fixed, nested in colony)] was applied. To test for additional differences in the quantity of CHCs, absolute concentrations (per surface area) were analyzed in the same way.

A multivariate approach was used to analyze relative changes in CHC composition (Bray-Curtis similarities) and the presence or absence of compounds, the latter evaluated by calculating resemblances based on "simple matching". A PERMANOVA with a 2-factor nested design, as described above, was applied for both analyses. Chromatograms of chemical profiles of host ants, spiders and silverfish can be found in Witte et al. (2009).

Behavioral Analysis

Aggression indices of isolated and non-isolated spiders were compared using a PERMANOVA, with a two-factor nested design, as described above. The interactions of spiders with host ants were evaluated in a multivariate approach, including all observed behaviors. They were standardized by total and a 2-factor nested design was applied as described above.

Comparison between Silverfish and Spiders

Since spiders and silverfish mostly did not originate from the same colonies, we did not consider colony as a factor for comparison. Accordingly, a PERMANOVA with a 1-factor design [species (fixed)] was applied.

Results

Chemical Transfer Experiment

In line with our expectations, workers extracted directly after the labeling treatment carried high concentrations of eicosane-D_{42} (median=108.87 ng/mm^2; Fig. 2). After the 3 d experimental phase, ant workers (adults and callows) still had the highest eicosane-D_{42} concentrations (median=31.62 ng/mm^2), followed by silverfish (median=7.48 ng/mm^2) and spiders (median=3.40 ng/mm^2). The concentration of eicosane-D_{42} was higher in silverfish than in spiders (PERMANOVA, $P=$

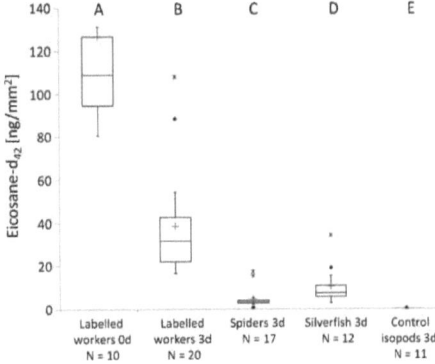

Fig. 2 Concentrations of eicosane-D_{42} in the chemical transfer experiment. Two outliers of labeled workers that were extracted directly after the labeling procedure (0 d) are not shown, allowing for better visibility (worker$_1$=241.51 ng/mm^2; worker$_2$=203.76 ng/mm^2). Different capital letters show differences (P<0.05) between groups evaluated by a PERMANOVA. Median (+=mean), quartiles (*boxes*), 10% and 90% percentiles (*whiskers*), and outliers (♦=outlier, *=extreme point) are shown. Abbreviations: 0 d=animals extracted directly after labeling; 3 d=animals extracted after the 3 day experimental phase

0.003). Importantly, control isopods had lower eicosane-D_{42} concentrations (median=0 ng/mm^2) than all other animals (PERMANOVA, for all pair-wise comparisons $P<0.001$).

Comparison of CHCs between Isolated and Non-isolated Spiders

For every single CHC, the concentration was lower in spiders after isolation. In addition, the total number of detected compounds decreased after isolation (across colonies: compounds of non-isolated spiders (Sp0d)=35, N=42; compounds of 9 d-isolated spiders (Sp9d)=8, N=25). There were 78 compounds above detection threshold on individual, non-isolated workers (N=49). No spider-specific compound was found.

Non-isolated spiders differed from isolated spiders in CHC composition, in the presence/absence of CHCs and in total CHC concentration (PERMANOVA, for all colonies $P\le0.015$; Table 3). Nine-days-isolated spiders had lower CHC concentrations than non-isolated spiders (medians across colonies: Sp0d=4.00 ng/mm^2, N=42; Sp9d= 0.16 ng/mm^2, N=25; Fig. 3). Workers had about 30 times greater concentrations of CHCs than non-isolated spiders, and about 700 times higher concentrations of CHCs than isolated spiders (median=112.15 ng/mm^2, N=49; Fig. 1 in supplement). Accordingly, workers had greater concentrations than both spider groups (PERMANOVA; $P<0.001$). Non-isolated spiders were chemically closer to host workers than were isolated individuals (Fig. 4).

Social Acceptance Experiment

Unmanipulated spiders were rarely attacked by ants, for all tested colonies (median$_{AI}$=0 for all colonies; for sample sizes see Table 1). The AIs of unmanipulated spiders did not differ among colonies (PERMANOVA, P=0.232).

Combined Experiment

Although isolated spiders of colony 4 showed reduced chemical resemblance to host workers (see above), they were rarely

Table 3 Comparison of non-isolated and isolated spiders for cuticular hydrocarbon (CHC) composition, presence or absence of CHCs, and total CHC concentration

Colony	CHC composition	CHC presence/absence	CHC concentration
Colony 1	0.001	0.003	0.001
Colony 2	0.003	0.015	0.001
Colony 3	0.001	0.001	0.001
Colony 4	0.001	0.001	0.001

PERMANOVA P values are shown. For sample sizes see Table 1

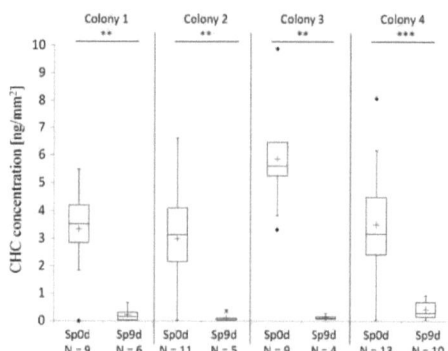

Fig. 3 Cuticular hydrocarbon (CHC) concentrations of non-isolated and 9 d-isolated spiders within colonies. Note that host colonies can differ in CHC concentrations and, therefore, CHC concentrations of spiders may differ as well. Two outliers among non-isolated spiders of colony 4 are not shown for better visibility (15.83 ng/mm^2 and 36.13 ng/mm^2). Median (+=mean), quartiles (boxes), 90% and 10% percentiles (whiskers), and outliers (♦=outlier, *=extreme point) are shown. Differences between groups were evaluated by a PERMANOVA (**$P<0.01$, ***$P\le0.001$). Abbreviations: Sp0d=non-isolated spiders; Sp9d=nine day-isolated spiders

attacked (median$_{AI}$=0; Fig. 5). Their AIs did not differ from those of non-isolated spiders (PERMANOVA, P=0.787). Considering all behavioral categories, we found no difference between isolated and non-isolated spiders (PERMANOVA; P=0.142). However, the low sample size does not allow us to exclude minor differences. When looking at each behavioral

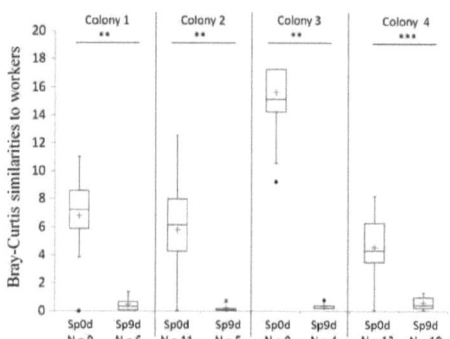

Fig. 4 Bray-Curtis similarities of individual spiders to the average chemical worker profile (N$_{workers}\ge10$) of their native host colony. One outlier of colony 3 is not shown for better visibility (value=25.21). Differences between groups were evaluated by a PERMANOVA (**$P<0.01$, *** $P<0.001$). Median (+=mean), quartiles (boxes), 10% and 90% percentiles (whiskers), and outliers (♦) are shown. Abbreviations: Sp0d=non-isolated spiders; Sp9d=nine days isolated spiders

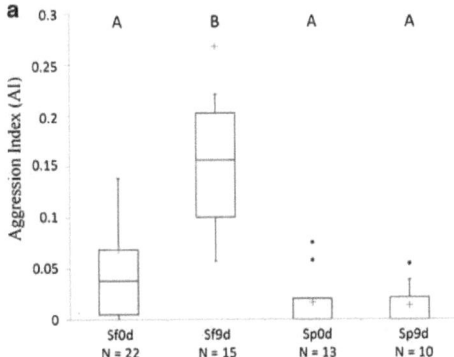
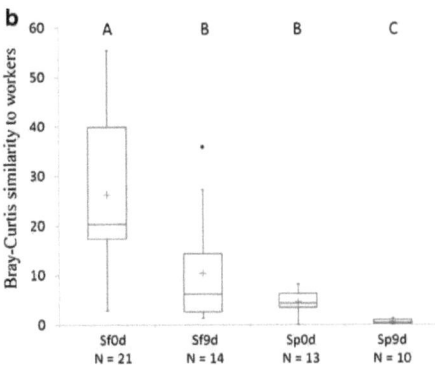

Fig. 5 Aggression indices of two myrmecophiles of *Leptogenys distinguenda*, the silverfish *Malayatelura ponerophila*, and the spider *Gamasomorpha maschwitzi* (a), and their chemical (Bray-Curtis) similarities to the average worker profile (b). Data originated from the same colony (colony 4). Three outliers of 9 d-isolated silverfish are not shown in the left graph for better visibility (AI=0.5 and two times AI=1). Median (+=mean), quartiles (*boxes*), 90% and 10% percentiles (whiskers), and outliers (♦= outlier, *=extreme point) are shown. Different capital letters depict differences ($P<0.05$) among groups evaluated by PERMANOVA. Abbreviations: Sf0d=non-isolated silverfish, Sf9d=9 d-isolated silverfish, Sp0d= non-isolated spiders; Sp9d=nine day-isolated spiders

category separately, we found a difference in antennation; i.e., isolated spiders were antennated by workers more frequently than non-isolated spiders (supplement Table 2). In addition, ant workers ignored non-isolated spiders more frequently than isolated spiders. For full details of each behavioral category, see supplement Table 2.

Comparison between the Spider and the Silverfish

Non-isolated individuals of both myrmecophilous species were attacked infrequently (silverfish: median$_{AI}$=0.02, N=67; spider: median$_{AI}$=0, N=41). However, we found differences in the social integration of both myrmecophiles, i.e., non-isolated silverfish were attacked more often in the social-acceptance experiment than were non-isolated spiders (PERMANOVA, P=0.003). Interestingly, non-isolated silverfish showed higher chemical similarities to host workers (median=50.85, N=51) than did non-isolated spiders (median=6.81, N=42; PERMANOVA, $P<0.001$). Furthermore, the concentrations of CHCs per surface area were higher in silverfish than in spiders (PERMANOVA; for all comparisons $P<0.001$; supplement Fig. 1).

Individuals of silverfish and spiders also were tested for the same host colony (colony 4). The AIs of non-isolated silverfish, isolated and non-isolated spiders did not differ (PERMANOVA, for all pairwise comparisons $P\geq0.113$), but the AIs of isolated silverfish were higher (PERMANOVA, $P\leq0.003$; Fig. 5). Non-isolated silverfish showed the highest chemical similarity to host workers (median=17.97), followed by isolated silverfish (median=5.40), non-isolated spiders (median=4.30), and isolated spiders (median=0.44;

Fig. 5). Notably, isolated silverfish and non-isolated spiders did not differ in chemical similarity to host workers (PERMANOVA, P=0.203).

Discussion

The spider *G. maschwitzi* acquired a chemical label from its host ants and showed reduced chemical resemblance to the host after isolation. Both results indicate that the myrmecophilous spider acquires mimetic CHCs rather than biosynthesizing them. In addition, clear differences between the social integration mechanisms of *G. maschwitzi* and the previously studied silverfish *M. ponerophila* were found. Contrary to expectation, spiders remained socially integrated in host colonies, in spite of experimentally reduced chemical host resemblance. Although spiders apparently do not depend as much as silverfish on high chemical host resemblance, the interactions between spiders and ants suggest that they may benefit from chemical mimicry (see below).

Origin of Mimetic CHCs The transfer of the stable-isotope label to spiders, but not to control isopods, demonstrates that the spiders acquire CHCs from the cuticle of host ants, probably through frequent body contact (Witte et al., 2009; see also video Online Resource 1). During these contacts, spiders often rub their legs, first on ant workers, and then on their own body (Witte et al., 2009) or through their own mouthparts (CvB, personal observation). Further evidence for behavioral acquisition of CHCs is demonstrated by isolation. While the CHC concentration of host workers did not decrease in isolation (see von Beeren et al., 2011b), the concentrations of mimetic CHCs, as well as qualitative chemical host resemblance, decreased in spiders when contact to host ants was prevented by isolation. The loss of host-specific compounds on the cuticle of a myrmecophile, when separated from its host, has also been shown in several other studies (Vander Meer and Wojcik, 1982; Vander Meer et al., 1989; Akino, 2008; von Beeren et al., 2011b). These results point to an acquisition of mimetic CHCs through host contact ('acquired chemical mimicry', sensu von Beeren et al., 2012) rather than through biosynthesis ('innate chemical mimicry', sensu von Beeren et al., 2012). However, the ability of the spiders to downregulate biosynthesis of mimetic CHCs in the absence of host ants cannot be fully excluded.

Acquisition of CHCs through contact with the host (e.g., rubbing behavior) seems to be a common strategy to facilitate integration (Lenoir et al., 2001), and has been suggested to exist in socially parasitic ants (Lenoir et al., 1997; Bauer et al., 2010), as well as other myrmecophiles (Vander Meer and Wojcik, 1982; Vander Meer et al., 1989). Acquisition, rather than innate biosynthesis of the complex host CHC profile, appears to be a more evolutionarily parsimonious mechanism for taxonomically distant myrmecophiles (von Beeren et al., 2011b). Nevertheless, the time during which previously isolated G. maschwitzi individuals interacted with host ants in the social acceptance experiment (median=8 min) was apparently insufficient to re-acquire an amount of CHC similar to that of non-isolated individuals.

The Role of Accuracy in Chemical Mimicry Previous studies on the multi-parasite system of the host ant *L. distinguenda* revealed that certain myrmecophiles are aggressively expelled (Witte et al., 2008; von Beeren et al., 2011a). This raises the question of how other myrmecophiles achieve integration without being attacked and killed. If nestmate recognition in ants is based mainly on a good match of colony-specific CHCs, as commonly seems to be the case (van Zweden and d'Ettorre, 2010), then myrmecophiles should either be driven toward chemical resemblance of hosts ('chemical mimicry', sensu von Beeren et al., 2012) or, alternatively, suppress expression of chemicals able to be detected by the host ('chemical hiding', sensu von Beeren et al., 2012), thus circumventing recognition as aliens. Although all compounds on the cuticles of ants potentially could be involved in recognition, we focused on non-polar compounds, because of the generally accepted role of CHCs in ant nestmate recognition. As hydrocarbons were abundant on the cuticles of the spider, and there were only traces of other compounds (supplement Table 1), we assume that the increased inspection of isolated spiders by host ants (see below) was caused predominantly by a mismatch in CHC signature. Evidence for reliance on chemical mimicry of host CHCs recently was found in the myrmecophilous silverfish *M. ponerophila* (von Beeren et al.,

2011b). We expected similar results for *G. maschwitzi*; i.e., a lower resemblance of host CHCs should result in increased recognition and ant rejection. Contrary to our expectation, isolated spiders were not attacked more frequently than unmanipulated spiders, regardless of their reduced accuracy in chemical host resemblance, which was manifest by lower concentrations of CHCs, absence of certain CHCs, and increased chemical distance.

A suppression of chemical recognition cues can be an adaptive strategy of parasites that prevents chemical detection by their hosts ('chemical hiding', sensu von Beeren et al., 2012). Chemical hiding, also referred to as chemical insignificance, has been suggested in various parasites of social insects (Lenoir et al., 2001; Lambardi et al., 2007; Nash and Boomsma, 2008; Baer et al., 2009; Kroiss et al., 2009). Due to the low levels of CHCs on the cuticle of *G. maschwitzi*, chemical hiding could be inferred. However, the spider's chemical appearance likely conveys information to the host, as ant behaviors differed to spiders showing either high or low accuracy of chemical host resemblance. Spiders were "ignored" less often, and "antennated" more often, by ant workers after isolation. Hence, it seems reasonable that workers detect spiders chemically and, therefore, chemical hiding does not apply. Instead, we argue that chemical host resemblance of spiders is best considered as mimicry (sensu von Beeren et al., 2012) because spiders most likely "deceive" the host by chemically resembling an entity of interest, specifically a host worker. This type of deception is likely beneficial for the spider for the following reason. As ant antennae carry mechano- and chemo-receptors, antennation may be a form of inspection that precedes subsequent reactions, including aggression (CvB, personal observation). We assume that unmanipulated spiders reduce inspection through "antennation" because of higher chemical resemblance, which thereby reduces the likelihood of subsequent attacks. Indeed, host attacks against spiders occur occasionally (Fig. 1C; also Witte et al., 2009). As isolated spiders were inspected more intensely, it remains to be explained why they were not attacked subsequently.

An evolutionary explanation for the low levels of host aggression against spiders may be a lower selection pressure on the host to recognize and fend off spiders, compared to silverfish, because of differences in virulence between the two myrmecophiles. Both species are kleptoparasites and do not prey on the host (Witte et al., 2008, 2009). Furthermore, there is no evidence of any benefits to the host from the presence of the two myrmecophiles. That both species are occasionally attacked and killed, points to a parasitic relationship, even if there is little impact on host fitness. Differences in virulence of the two kleptoparasites may arise from the amount of host resources they consume. While body sizes are comparable, silverfish occur in higher numbers (about 4–5 times that of spiders; Witte et al., 2008).

Consequently, silverfish may have a greater negative impact on the host. There is evidence that ants can direct their defense specifically against more costly parasites (von Beeren et al., 2011a). Accordingly, lower aggression against spiders might be explained by lower selection on the host to recognize and fend-off spiders.

Social Integration Mechanisms From the perspective of parasites, decreased host defense against spiders, compared to silverfish, suggests that spiders have evolved additional integration mechanisms that achieve and maintain social integration. Two main differences are obvious between *G. maschwitzi* and *M. ponerophila*, one behavioral and the other morphological. The spider's response to host aggression differs from that of the silverfish. Silverfish attempt to escape, whereas spiders remain stationary until ant aggression ceases (Witte et al., 2009). The stationary behavior apparently causes fewer attacks by the ants, perhaps because escape is a typical behavior of prey items. In laboratory colonies, spiders moved freely among host ants, frequently interacting directly with host workers through their long, thin legs (Witte et al., 1999, 2009). Spiders constantly adjust their position (CvB, personal observation), often resulting in them sitting on top of ants (Witte et al., 2009; see also video of supplement). This may further help spiders avoid confrontation with ant workers. Additionally, our impression is that the movements of spiders resemble those of the ants (see also video Online Resource 1), whereas silverfish appear to move quite differently. This behavioral similarity may make tactile recognition of the spiders as alien difficult. We suspect that these behaviors facilitate peaceful interactions between spiders and host ants.

The formicoid habitus (morphological resemblance of ants) of certain myrmecophiles is likely an adaptation directed at predators (Hölldobler and Wilson, 1990; Nelson, 2011). However, certain morphological traits of myrmecophiles, such as cuticular surface structure, also may serve as an ancillary integration mechanism for host deception (Hölldobler and Wilson, 1990). Tactile mimicry, also called Wasmannian mimicry after its first description (Wasmann, 1895), means that a worker's tactile inspection cannot distinguish the body constitution and the surface structure between a mimic and its model (Gotwald, 1995). Myrmecophilous mites of the genus *Planodiscus*, for example, resemble the surface structure of their host to a high degree (Kistner, 1979). We hypothesize that the spider's body shape and surface, which is covered with setae, sufficiently resembles the host's body, while the silverfish have a completely different body constitution (Fig. 2 supplement). The limuloid, scaled body of the silverfish, with short appendices (antennae, cerci and praecerci) and retractable head, are probably adaptations to escape from ant attacks, rather than adaptations to interact peaceably with them. As spiders mainly interact with workers using their thin, long, setae-covered legs, we suspect that the legs play an important role in deceiving the host, with workers misidentifying spiders as nestmates. Ants frequently groom nestmates (Hölldobler and Carlin, 1989), and grooming behavior can also be observed directed toward spiders (see Table 2 of supplement). Given that the myrmecophile's body parts that frequently interact with the host carry potential recognition cues that need to be mimicked for favorable recognition (Gotwald, 1995), we hypothesize a central role for the spider's legs.

Transmission between Colonies The spider's lower reliance on chemical mimicry also may be beneficial for their dispersal. Invading new host colonies (horizontal transmission) is a difficult task for myrmecophiles that rely on chemical integration mechanisms, because ant colonies possess a colony-specific odor (van Zweden and d'Ettorre, 2010). Since high accuracy of chemical mimicry is necessary for social acceptance in the silverfish, this myrmecophile is likely to face greater difficulties in transmitting horizontally between different colonies than will the spiders. Indeed, spiders have been exchanged successfully between *L. distinguenda* colonies with little or no increased aggression, whereas silverfish were always killed during exchange experiments (Witte et al., 2009). Since new *L. distinguenda* colonies most probably bud from old nests, as is the case in other army ants (Kronauer, 2009), silverfish most likely are mainly limited to vertical transmission from mother to daughter colonies.

In summary, combinations of several integration mechanisms, e.g., chemical strategies, acoustical, behavioral, and morphological adaptations, allow myrmecophiles to integrate with their host ants. As demonstrated here, the degree of dependency on some of these mechanisms may differ between myrmecophilous species. Thus, the more integration mechanisms are studied in combination, the more reliably results will demonstrate which adaptations are most important in the social integration of myrmecophiles.

Acknowledgements We thank the behavioral ecology group at the LMU Munich for helpful comments on the manuscript, with special thanks to Sebastian Pohl and Andrew Bruce. Many thanks also to Sofia Lizon à l'Allemand, Max Kölbl, Magdalena Mair, and Deborah Schweinfest for assistance in the field. We are grateful for financial support from the DFG (Deutsche Forschungsgemeinschaft, project WI 2646/3).

References

AKINO, T. 2008. Chemical strategies to deal with ants: a review of mimicry, camouflage, propaganda, and phytomimesis by ants (Hymenoptera:Formicidae) and other arthropods. *Myrmecol. News* 11:173–181

ALLAN, R. A., CAPON, R. J., BROWN, W.V., and ELGAR, M.A. 2002. Mimicry of host cuticular hydrocarbons by salticid spider

Cosmophasis bitaeniata that preys on larvae of tree ants *Oecophylla smaragdina*. *J. Chem. Ecol.* 28:835–848

ANDERSON, M. J., GORLEY, R. N., and CLARKE, K. R. 2008. PERMANOVA+ for PRIMER: Guide to software and statistical methods. PRIMER-E, Plymouth, UK

BAER, B., DEN BOER, S. P. A., KRONAUER, D. J. C., NASH, D.R., and BOOMSMA, J. J. 2009. Fungus gardens of the leafcutter ant *Atta colombica* function as egg nurseries for the snake *Leptodeira annulata*. *Insectes Soc.* 56:289–291

BAGNÈRES, A.-G., and LORENZI, M. 2010. Chemical deception/mimicry using cuticular hydrocarbons. *in*: Blomquist GJ, Bagnères A-G (eds) Insect hydrocarbons: Biology, Biochemistry and Chemical ecology. Cambridge University Press, New York, USA

BATES, H.W. 1862. Contributions to an insect fauna of the Amazon Valley. Lepidoptera: Heliconidae. *Trans. Linn. Soc.* 23:495–566

BAUER, S., BOEHM, M., WITTE, V., and FOITZIK, S. 2010. An ant social parasite in-between two chemical disparate host species. *Evol. Ecol.* 24:317–332

BROWER, L. P. 1988. Mimicry and the evolutionary process. *Amer. Nat.* 131: Supplement S1-S121

CECCARELLI, F. S. 2007. Contact between *Myrmarache* (Araneae: Salticidae) and ants. *Bull. Br. Arachnol. Soc.* 14:54–58

CUSHING, P. E. 1997. Myrmecomorphy and myrmecophily in spiders: a review. *Florida Entomol.* 80:165–193

DETTNER, K. and LIEPERT, C. 1994. Chemical mimicry and camouflage. *Annu. Rev. Entomol.* 39:129–154

EMERY, C. 1887. Catalogo delle formiche esistenti nelle collezioni del Museo Civico. *Gen Insect Fasc* 118:1–124

FISHER, R. A. 1927. On some objections to mimicry theory: statistical and genetic. *Trans. Ent. Soc. London* 75:269–278

GOTWALD, W. H. JR. 1995. Army Ants: The Biology of Social Predation. Cornell University Press, New York

HÖLLDOBLER, B. and CARLIN, N. F. 1989. Colony founding, queen control and worker reproduction in the ant *Aphaenogaster* (=*Novomessor*) *cockerelli* (Hymenoptera: Formicidae). *Psyche* 96:131–151

HÖLLDOBLER, B. and WILSON, E.O. 1990. The Ants. Harvard University Press, Cambridge

KISTNER, D. H. 1979. Social and evolutionary significance of social insect symbionts. pp 339–413 *in*: Hermann HR (ed) Social insects. Academic Press, New York.

KROISS, J., SCHMITT, T., and STROHM, E. 2009. Low level of cuticular hydrocarbons in a parasitoid of a solitary digger wasp and its potential for concealment. *J. Entomol. Sci.* 12:9–16

KRONAUER, D. J. C. 2009. Recent advances in army ant biology (Hymenoptera: Formicidae). *Myrmecol. News* 12:51–65

LAMBARDI, D., DANI, F.R., TURILLAZZI, S., and BOOMSMA, J. J. 2007. Chemical mimicry in an incipient leaf-cutting ant social parasite. *Behav. Ecol. Sociobiol.* 61:843–851

LENOIR, A., D'ETTORRE, P., ERRARD, C., and HEFETZ, A. 2001. Chemical ecology and social parasitism in ants. *Annu. Rev. Entomol.* 46:573–599

LENOIR, A., MALOSSE, C., and YAMAOKA, R. 1997. Chemical mimicry between parasitic ants of the genus *Formicoxenus* and their host *Myrmica* (Hymenoptera, Formicidae). *Biochem. Syst. Ecol.* 25:379–389

MENDES, V., VON BEEREN, C., and WITTE, V. 2011. *Malayatelura ponerophila* - a new genus and species of silverfish (Zygentoma, Insecta) from Malaysia, living in *Leptogenys* army-ant colonies (Formicidae). *Dtsch. Ent. Z.* 58:193–200

MÜLLER, J. 1878. Über die Vortheile der Mimikry bei Schmetterlingen. *Zoolog. Anz.* 1: 54–55

NASH, D. R., ALS, T. D., MAILE, R., JONES, G. R., and BOOMSMA, J. J. 2008. A mosaic of chemical coevolution in a large blue butterfly. *Science* 319:88–90

NASH, D. R., and BOOMSMA, J. J. 2008. Communiction between hosts and social parasites. *in*: d'Ettorre P, Hughes DP (eds) Sociobiology of communication an interdisciplinary approach. Oxford University Press, New York

NELSON, X. J. 2011. A predator's perspective of the accuracy of ant mimicry in spiders. *Psyche* 2012; doi:10.1155/2012/168549

RUXTON, G. D., SHERATT, T., and SPEED, M. 2004. Avoiding Attack: The Evolutionary Ecology of Crypsis, Warning Signals and Mimicry. Oxford University Press, New York

STEGHAUS-KOVAC, S. 1994. Wanderjäger im Regenwald- Lebensstrategien im Vergleich: Ökologie und Verhalten südostasiatischer Ameisenarten der Gattung *Leptogenys* (Hymenoptera: Formicidae: Ponerinae). Dissertation, University Frankfurt, Germany

STEIGER, S., SCHMITT, T., and SCHÄFER, H. M. 2011. The origin and dynamic evolution of chemical information transfer. *Proc. R. Soc. B.* 278:970–979

STEIN, S. E. 1999. An Integrated Method for Spectrum Extraction and Compound Identification from GC/MS Data. *J. Am. Soc. Mass Spectrom.* 10:770–781

SYMONDS, M. R. E. and ELGAR, M. A. 2008. The evolution of pheromone diversity. *Trends Ecol. Evol.* 23:220–228

TANNER, C. J. and ADLER, F. R. 2009. To fight or not to fight: context-dependent interspecific aggression in competing ants. *Anim. Behav.* 77:297–305

VAN ZWEDEN, J. S. and D'ETTORRE, P. 2010. Nestmate recognition in social insects and the role of hydrocarbons. pp 222–243 *in*: Blomquist GJ, Bagnères AG (eds) Insect hydrocarbons: Biology, Biochemistry and Chemical Ecology. Cambridge University Press, New York.

VANDER MEER, R.K., JOUVENAZ, D. P., and WOJCIK, D. P. 1989. Chemical mimicry in a parasitoid (Hymenoptera: Eucharitidae) of fire ants (Hymenoptera: Formicidae). *J. Chem. Ecol.* 15:2247–2261

VANDER MEER, R.K. and WOJCIK, D. P. 1982. Chemical mimicry in the myrmecophilous beetle *Myrmecaphodius excavaticollis*. *Science* 218:806–808

VON BEEREN, C., MARUYAMA, M., HASHIM, R., and WITTE, V. 2011a. Differential host defense against multiple parasites in ants. *Evol. Ecol.* 25:259–276

VON BEEREN, C., POHL, S., and WITTE, V. 2012. On the use of adaptive resemblance terms in chemical ecology. *Psyche*, Article ID 635761, doi:10.1155/2012/635761

VON BEEREN, C., SCHULZ, S., HASHIM, R., and WITTE, V. 2011b. Acquisition of chemical recognition cues facilitates integration into ant societies. *BMC Ecology* 11:30

WASMANN, E. 1895. Die Ameisen-und Termitengäste von Brasilien. I. Teil. Mit einem Anhange von Dr. August Forel. *Verh K K Zool Bot Ges Wien* 45:137–179

WICKLER, W. 1968. Mimikry. Nachahmung und Täuschung in der Natur. Kindlers Universitäts Bibliothek, München

WILSON, E. O. 1990. Success and dominance in ecosystems: the case of the social insects. Excellence in ecology, 2. Ecology Institute, Oldendorf/Luhe

WITTE, V., FOITZIK, S., HASHIM, R., MASCHWITZ, U., and SCHULZ, S. 2009. Fine tuning of social integration by two myrmecophiles of the ponerine army ant, *Leptogenys distinguenda*. *J. Chem. Ecol.* 35:355–367

WITTE, V., HÄNEL, H., WEISSFLOG, A., HASHIM, R., and MASCHWITZ, U. 1999. Social integration of the myrmecophilic spider *Gamasomorpha maschwitzi* (Araneae: Oonopidae) in colonies of the South East Asian army ant, *Leptogenys distinguenda* (Formicidae: Ponerinae). *Sociobiology* 34:145–159

WITTE, V., LEINGÄRTNER, A., SABAß, L., HASHIM, R., and FOITZIK, S. 2008. Symbiont microcosm in an ant society and the diversity of interspecific interactions. *Anim. Behav.* 76:1477–1486

WUNDERLICH, J. 1994. Beschreibung bisher unbekannter Spinnenarten und -gattungen aus Malaysia und Indonesien (Arachnida: Araneae: Oonopidae, Tetrablemmidae, Telemidae, Pholcidae, Linyphiidae, Nesticidae, Theridiidae und Dictynidae). *Beiträge Araneologie* 4:559–579

Supplementary Material

Title: Social integration of a myrmecophilous spider does not depend exclusively on chemical mimicry

Christoph von Beeren, Rosli Hashim and Volker Witte

Calculation of spider surface areas

The surface area of the pro- and opistosoma of spiders was calculated by applying the formula of a three-axial ellipsoid according to Thomson[1]:

$$\text{Surface area}_{\text{ellipsoid}} = 4\pi \left[\frac{(ab)^{1.6} + (ac)^{1.6} + (ab)^{1.6}}{3} \right]^{0.625}$$

The variables a, b and c refer to the length, the breadth and the height of the ellipsoid. The surface area of the spiders' coxa, trochanter, femur, tibia, tarsus of each leg was calculated separately by the formula of the lateral surface area of a circular cylinder:

$$\text{Surface}_{\text{lateral area of circular cylinder}} = 2\pi rh = dh\pi$$

The variables h and d refers to the length and breadth of the respective part of the spiders' leg. Finally, the surface areas of all parts were summed up. Specimens were stored in 99% ethanol.

[1]: http://www.numericana.com/answer/ellipsoid.htm#thomsen

Table 1. Concentrations of the 32 compounds that contributed to 99.08 % of the chemical profiles of workers across colonies, evaluated by a similarity percentage analysis (SIMPER) on Bray-Curtis distances. Structural alkene isomers are marked using different capital letters. Abbreviations: Sp 0 d = non-isolated spiders; Sp 9 d = nine days isolated spiders; SE = standard error

Compound	Concentrations [µg/mm^2 ± SE]		
	Workers (N = 49)	Sp 0 d (N = 42)	Sp 9 d (N = 25)
Nonacosene (A)	17.950 ± 1.151	0.479 ± 0.086	0.009 ± 0.006
Tricosane	15.873 ± 1.028	1.984 ± 0.125	0.182 ± 0.037
Heptacosene (A)	12.348 ± 1.037	0.262 ± 0.043	0.001 ± 0.001
Pentacosene (A)	11.006 ± 0.999	0.359 ± 0.060	0.005 ± 0.004
Pentacosadiene	8.736 ± 0.554	0.110 ± 0.031	-
Pentacosene (B)	5.625 ± 0.384	0.151 ± 0.028	-
Hentriacontene	6.000 ± 0.556	0.004 ± 0.003	-
Heptacosadien	4.791 ± 0.476	-	-
Pentacosane	3.206 ± 0.211	0.309 ± 0.036	0.069 ± 0.014
Heptacosene (B)	3.517 ± 0.505	0.005 ± 0.003	-
Nonacosene (B)	2.285 ± 0.179	-	-
Pentacosene (C)	1.505 ± 0.114	0.009 ± 0.005	-
3,6,9- Pentacosatriene	1.682 ± 0.252	0.004 ± 0.004	-
Nonacosadiene	1.756 ± 0.291	-	-
11-Methylpentacosane	0.959 ± 0.090	0.002 ± 0.002	-
Tricosene (B)	0.828 ± 0.072	-	-
11-Methylheptacosane	0.798 ± 0.061	0.005 ± 0.005	-
Tricosene (C)	1.119 ± 0.180	0.028 ± 0.010	-
Tetracosane	0.642 ± 0.042	0.022 ± 0.006	0.001 ± 0.000
Decyloctanoate	0.774 ± 0.095	-	-
Tritriacontene	0.802 ± 0.115	-	-
Decyl decanoate	0.540 ± 0.069	-	-
Docosane	0.282 ± 0.020	0.052 ± 0.034	-
13- and 15- Methylnonacosanes	0.394 ± 0.050	-	-
13- and 15- Methylhentriacontanes	0.435 ± 0.078	-	-
6,9,12,15-Heptacosatetraene	0.342 ± 0.053	-	-
Octacosene (A)	0.375 ± 0.067	-	-
Tricosene (A)	0.322 ± 0.067	-	-
Multiple methylated hentriacontenes	0.441 ± 0.094	-	-
Heneicosane	0.159 ± 0.019	0.029 ± 0.017	-
9-Methyltricosane	0.120 ± 0.011	-	-
Heptacosane	0.121 ± 0.012	0.027 ± 0.019	-

Table 2. Behavioural interactions between ants and spiders that were either not isolated (Sp0d) or isolated for nine days (Sp9d) from their host (colony 4).

Interaction	Sp0d N = 13		Sp9d N = 10	
Ignored	158 / 15	A	84 / 9	B
Groomed	108 / 8	A	102 / 11.5	A
Avoid	65 / 3	A	41 / 2	A
Unnoticed	216 / 16	A	140 / 12	A
Antennated	105 / 7	A	151 / 13	B
Snapped	4 / 0	A	2 / 0	A
Chased	4 / 0	A	5 / 0	A
Stung	3 / 0	A	2 / 0	A

The upper number in each cell represents the sum and the lower number the median of the corresponding interaction. Different capital letters depict significant differences ($p < 0.05$) between isolation treatments for a given behavioural interaction (row), evaluated by a PERMANOVA. Abbreviations: Sp0d = non-isolated spiders; Sp9d = nine days isolated spiders.

Figure 1. CHC concentration (quantity per surface area) of silverfish ($N_{colonies}$ = 4), spiders ($N_{colonies}$ = 4) and ant workers ($N_{colonies}$ = 5). Different capital letters depict significant differences ($p < 0.001$) between groups evaluated by a PERMANOVA. Abbrevistions: Sf0d = non-isolated silverfish; Sf9d = nine days isolated ssilverfish; Sp0d = non-isolated spiders; Sp9d = nine days isolated spiders

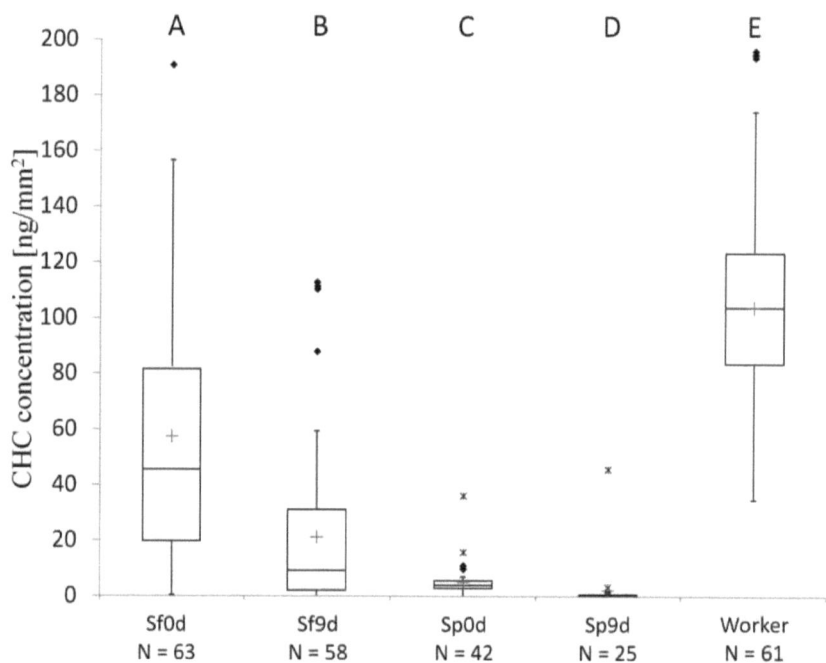

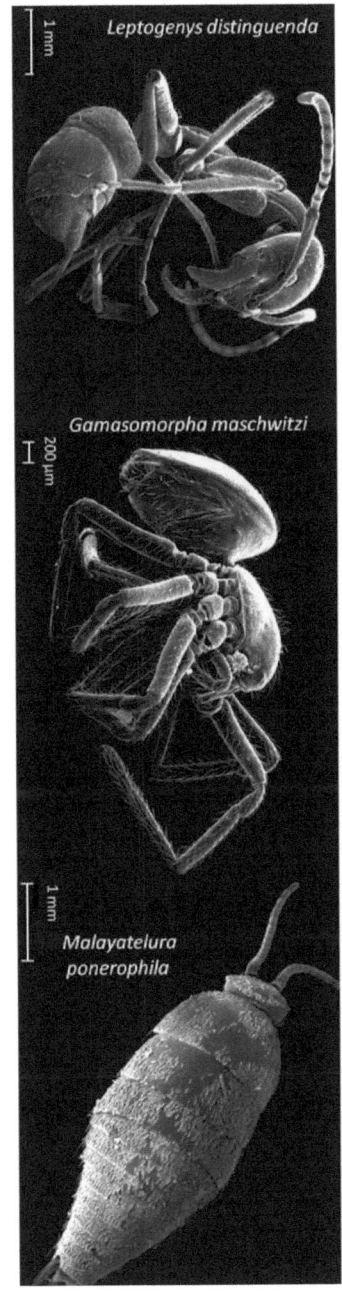

Figure 2. We assume that the spider *G. maschwitzi* resembles the body constitution and the hairy surface of the host better than the silverfish *M. ponerophila*, which has a limuloid shaped body with scales on its surface. © Max Kölbl

Video. A video is available on the following website: http://www.springerlink.com/content/m25j76j00h243q61/

The video (Online Resource 1) shows typical spider behaviours in a laboratory nest. *Gamasomorpha maschwitzi* often sits on top of ant workers (see also (Witte et al. 2009)). By rubbing its legs on ant bodies, the spider presumably acquires host CHCs. Spiders are able to follow the ants in a "tandem running"-like behaviour, and are thereby able to follow the host during migrations. This video is a cut-out from a documentary on various myrmecophiles of the army ant *L. distinguenda*. The whole movie can be downloaded on the following webpage:

http://www.evolution-of-life.com/en/observe/video/fiche/an-evolutionary-arms-race.html

Chapter 4

On the use of adaptive resemblance terminology in chemical ecology

Christoph von Beeren, Sebastian Pohl, and Volker Witte

In adaptive resemblance the mimic modifies its appearance, pretending to be something different, in order to dupe another organism. This spider likely performs 'crypsis' by matching background cues to dupe predators such as birds. © Christoph von Beeren

2012

◆

von Beeren C, Pohl S, and Witte V, in press (online first). On the use of adaptive resemblance terminology in chemical ecology. *Psyche*, doi:10.1155/2012/635761

Review Article
On the Use of Adaptive Resemblance Terms in Chemical Ecology

Christoph von Beeren, Sebastian Pohl, and Volker Witte

Department of Biology II, Ludwig Maximilians University Munich, Großhaderner Straße 2, 82152 Planegg-Martinsried, Germany

Correspondence should be addressed to Volker Witte, witte@biologie.uni-muenchen.de

Received 12 September 2011; Revised 7 November 2011; Accepted 8 November 2011

Academic Editor: Alain Lenoir

Copyright © 2012 Christoph von Beeren et al. This is an open access article distributed under the Creative Commons Attribution License, which permits unrestricted use, distribution, and reproduction in any medium, provided the original work is properly cited.

Many organisms (mimics) show adaptive resemblance to an element of their environment (model) in order to dupe another organism (operator) for their own benefit. We noted that the terms for adaptive resemblance are used inconsistently within chemical ecology and with respect to the usage in general biology. Here we first describe how resemblance terms are used in general biology and then comparatively examine the use in chemical ecology. As a result we suggest the following consistent terminology: "chemical crypsis" occurs when the operator does not detect the mimic as a discrete entity (background matching). "Chemical masquerade" occurs when the operator detects the mimic but misidentifies it as an uninteresting entity, as opposed to "chemical mimicry" in which an organism is detected as an interesting entity by the operator. The additional terms "acquired" and "innate" may be used to specify the origins of mimetic cues.

1. Introduction

Social insects, especially ants and termites, dominate many terrestrial habitats in terms of abundance, biomass, and energy turnover [1, 2]. They accumulate considerable amounts of resources that can be of potential use for other organisms, in the form of living biomass, infrastructures (e.g., nest sites), or stored products [3]. The ecological success of social insects comes with the cost that predators and parasites may exploit their societies [4–6]. Since Wasmann's [7] extensive study on organisms that developed close relationships with ants, a multitude of so-called myrmecophiles has been found to exploit ant colonies and their social resources in a variety of ways [5, 8]. Parasitic relationships may escalate in an evolutionary arms race where the hosts adapt towards protecting themselves from exploitation, while parasites adapt towards avoiding expulsion from the host [9].

In this context it is crucial that members of a society can be recognized reliably and distinguished from aliens, which can thus be aggressively expelled [10]. An efficient social recognition system is essential for a colony to function as a closed unit. The better such recognition works, the more effectively social exploitation can be prevented. Complex profiles of cuticular hydrocarbons (CHCs) are known to carry information necessary for recognition of colony members in ants, bees, and wasps [10].

Macroparasites of ants have evolved a variety of strategies to cope with their hosts' elaborate recognition system [5]. Potential strategies for avoiding or resisting the hosts' defense behavior include the use of morphological, acoustical, and behavioral adaptations or the use of chemical repellents or attractants [1, 5, 11–13]. Particularly widespread and important are chemical strategies for avoiding recognition, either by not expressing relevant recognition cues or by matching host recognition cues [11, 14, 15]. For simplicity, we use the term "cue" referring to any chemical information that is potentially perceivable, irrespective of whether the information transfer is "intentional" or "unintentional" sensu Steiger et al. [16].

Chemical resemblances work analogously to other biological resemblances, such as acoustic or visual mimicry [17]. Unfortunately, different definitions exist in chemical ecology (see below), and thus different authors may describe different forms of chemical resemblances with identical terms or the same type of resemblance with different terms.

The aim of this paper is threefold. First, we identify how definitions of resemblances are generally used in biology. Second, we analyze the terminology that is used in chemical

TABLE 1: Summarized table of adaptive resemblance terms in general biology as used in important reviews. Systems can either be considered according to what a mimic pretends to be or according to what an operator perceives. We adopted the latter view.

	By an operator, the mimic is...		
not detected as a discrete entity (causing no reaction)	detected as an uninteresting entity (causing no reaction)	detected as an interesting entity (causing a reaction beneficial to the mimic)	Reference(s)
Crypsis	Masquerade	Mimicry	Endler 1981 [21], 1988 [22]
Eucrypsis	Mimesis	Homotypy	Pasteur 1982[a] [23]
Eucrypsis	Plant-part mimicry	Mimicry	Robinson 1981 [24]
Crypsis	Masquerade	Mimicry	Ruxton et al. 2004 [25], Ruxton 2009 [17]
Cryptic resemblance	Cryptic resemblance	Sematic resemblance	Starrett 1993 [18]
Crypsis	Masquerade	—	Stevens and Merilaita 2009[b] [26]
Crypsis	Crypsis	Mimicry[c]	Vane-Wright 1976 [27], 1980 [20]
Camouflage or mimesis	Camouflage or mimesis	Mimicry	Wickler 1968 [19]

—: not considered.
[a]Pasteur [23] uses the term "camouflage" as generic term for both eucrypsis and mimesis.
[b]The term "camouflage" is used by Stevens and Merilaita [26] to describe all forms of concealment, including crypsis and masquerade.
[c]For the imitation of inanimate objects, Vane-Wright [27] uses the expressions "decoys" or "deflective marks".

ecology. Finally, we attempt a synthesis and suggest a terminology that agrees best with the general biological definitions and with the chemical strategies observed in nature.

2. General Definitions of Biological Resemblances

Since the resemblance of organisms to elements of their environment (e.g., other organisms or background) is often not coincidental, but rather evolved for the benefit of the mimic, the term adaptive resemblance was coined [18]. In adaptive resemblance one organism (the mimic) modifies its appearance, pretending to be something different (the model), in order to dupe another organism (the operator) [19, 20]. Many different terms have been used to describe adaptive resemblance, including mimicry, camouflage, crypsis, masquerade, and mimesis. These terms have been debated intensively and defined repeatedly according to different criteria (see Table 1).

For the purpose of this paper, we adopted an operator's view to narrow down the existing definitions of adaptive resemblance into a unified system. This means that we distinguish the cues of a mimic with respect to whether and how they are perceived by the operator. The resulting categories are only valid within a given perceptive channel between the mimic and operator, and they can differ in other channels or if other organisms are considered. The first column of Table 1 defines resemblances in which a mimic is not perceived as a discrete entity by the operator and consequently causes no reaction in the operator. In such cases the mimic frequently blends with the background. We adopt the term "*crypsis*" for this phenomenon according to Endler [21], who first distinguished this type of resemblance from "*masquerade*". In the latter a mimic is perceived by an operator as a discrete entity, which is however misidentified as uninteresting so that the operator also shows no reaction to the mimic. Accordingly, crypsis relies on the relationship between the organism and the background, whereas the benefit of masquerade is thought to be independent of the background [28]. A stick insect, for example, is likely to be recognized as a stick by a potential predator independent of its surroundings (e.g., when lying on grass). A cryptic organism, however, depends strongly on the background. This fact allows testable predictions to be made. For example, a mimic performing masquerade should be treated similarly by the operator independent of its background. On the other hand a mimic that performs crypsis should be treated differently (e.g., recognized and attacked) by the operator when the background changes.

The third column of Table 1 defines adaptive resemblances in which a mimic is perceived by the operator as an entity of interest. This category was first described in a biological context by Bates [29] as "*mimicry*", and this term is currently most frequently used, hence we adopt it here.

Finally, another mechanism exists to avoid detection by an operator, which is however not based on resemblance. The term "*hiding*" has been applied to cases in which the absence of informative cues is achieved by behavioral adaptations, making detection by an operator impossible [17]. In visual systems, for example, a rabbit is hiding if it stays in its burrow in the presence of a predator (operator), thereby avoiding detection [17]. If a hiding organism would be removed from the environment, the perceptive input of the operator will not change in the concerning channel. Hiding is not included in Table 1 because it does not fall into categories of resemblance; nevertheless this term will be of importance in our discussion on chemical interactions below.

3. The Use of Adaptive Resemblance Terms in Chemical Ecology

Compared to visual adaptive resemblances, chemical adaptive resemblances had initially been paid less attention to in scientific literature, despite the fact that chemical communication is the most widespread form of communication among organisms [16, 30, 31]. However, more recent reviews on this topic show that understanding of chemical adaptive resemblance has increased markedly [11, 15, 32, 33].

According to this special issue on ants and their parasites, we focus here particularly on important reviews about parasites of social insects and on reviews about adaptive chemical resemblance. Reviews are suitable for analyzing how the terminology is used, since they provide overviews about specific fields, summarize the literature, and therefore mirror common practices.

We used the same categorization as in Table 1, adopting an operator's point of view. Note that two resemblance types were combined, that is, resemblances in which a mimic is not detected as discrete entity and resemblances in which a mimic is detected as an uninteresting entity (Table 2). We combined these two types of resemblances because none of the reviews distinguished them. Additionally, we included the origins of mimetic compounds in the table, since this is an interesting point regarding chemical resemblances and several authors based their terminology upon it.

Table 2 shows that the terms chemical mimicry and chemical camouflage are not used consistently. Some authors used the terms according to criteria similar to those used in general biology (see Table 1). They distinguished between chemical mimicry as the imitation of an interesting entity and chemical camouflage either as the imitation of an uninteresting entity or as the resemblance of background cues (sensu Dettner and Liepert [15]). This use of terms did not include the origins of mimetic compounds. In contrast, other authors focused primarily on the origin of mimetic cues. According to their terminology, chemical mimicry implies that mimetic cues are biosynthesized by the mimic, while chemical camouflage implies that the mimic acquires mimetic cues from the model (first defined by Howard et al. [38]). Additional definitions specifically focused on a mimic's avoidance of being detected as a discrete entity (Table 2). Chemical resemblances that allow mimics to avoid detection by background matching were defined as chemical mimesis by Akino [14] or as chemical crypsis by Stowe [31].

In addition to adaptive resemblances, another mechanism exists among parasites to prevent detection by an operator. This mechanism was called "chemical insignificance" [39]. However, chemical insignificance was originally brought up to describe the status of freshly hatched ant workers (callows), which typically carry very low quantities of cuticular hydrocarbons [39]. The term insignificance referred to these weak chemical cues, which are frequently not colony or even species specific, allowing the transfer and acceptance of callows into alien colonies [11]. The term chemical insignificance was also adopted to describe a status of ant parasites, which may benefit from displaying no or only small quantities of recognition cues to sneak unnoticed into host colonies [3, 11, 39, 40]. We discuss this point in more detail at the end of the following chapter.

Furthermore, chemical transparency was recently described as a chemical strategy in a wasp social parasite [41]. This strategy is somewhat similar to chemical insignificance, except that it refers particularly to a subset of cuticular compounds that are presumably responsible for recognition. We discuss both strategies, chemical insignificance and transparency, in more detail at the end of the following section.

4. Suggestions for a Consistent Terminology

As described above, adaptive resemblance terminology is used inconsistently in important reviews of chemical ecology, likely mirroring inconsistent use in this field generally. Most importantly, the terms chemical camouflage and chemical mimicry are inconsistently used by different approaches. While some authors distinguish them according to different models that are mimicked, others distinguish them according to the origin of mimetic cues (Table 2). To avoid confusion, we suggest a consistent terminology that is in line with the definitions used in general biology (Table 1). Consequently, adaptive resemblance of an entity interesting for the operator should be referred to as *"chemical mimicry"*, irrespective of the origin of mimetic cues. Nevertheless, an additional distinction between biosynthesis and acquisition of mimetic cues might often be useful. Hence, we suggest using additional terms to distinguish the origins of mimetic cues; *"acquired chemical mimicry"* indicates that mimetic cues are acquired from the model, while *"innate chemical mimicry"* (as first mentioned by Lenoir et al. [11]) indicates that a mimic has an inherited ability to biosynthesize mimetic compounds. The two different mechanisms may affect coevolutionary dynamics in different ways. For example, a consequence of the acquisition of recognition cues by a parasite from its host is that the mimetic cues of model and mimic are of identical origin [3]. Coevolutionary arms races select in such cases for effective ways of acquiring chemical host cues by the mimic, for example, through specific behaviors such as intensive physical contact to the host. In the host, selection favors counterdefenses which prevent the acquisition of chemical cues. Selection pressures are somewhat different when a parasite biosynthesizes mimetic cues [3]. In this case, the origins of the chemical cues of mimic and model are different, which allows coevolutionary arms races to shape on the one hand the accuracy of chemical mimicry of the mimic and on the other hand the discrimination abilities of the operator.

Mimics that are not detected as discrete entities or that are detected but misidentified as uninteresting entities by an operator have rarely been addressed in chemical ecological reviews, although they are common in general biology (first two columns of Table 1). Since the term camouflage is not used in general biology to distinguish these two forms of resemblances (Table 1) and since the term chemical camouflage is used inconsistently in chemical ecology (Table 2), we suggest abandoning this term so as to avoid confusion. Instead, we suggest using terms consistent to general biology. Accordingly, *"chemical crypsis"* describes cases in which an operator is not able to detect a mimic as a discrete entity, while *"chemical masquerade"* describes cases in which an operator detects a mimic as an uninteresting entity. In both cases, the operator shows no reaction. The terms "acquired" and "innate" can be applied to these categories as well to add further information on the origin of the disguising cues. Note that it is challenging but logically possible to

TABLE 2: Summarized table of the main terms used for chemical adaptive resemblances in reviews about parasites of social insects and in reviews about adaptive chemical resemblance. Systems can either be considered according to what a mimic pretends to be or according to what an operator perceives. We adopted the latter view. Furthermore, the terminology based on the origins of mimetic compounds is shown.

By an operator, the mimic is...		Origin of mimetic compounds in cases where the mimic is detected as interesting entity by the operator		Reference
not detected as discrete entity or detected as an uninteresting entity[a]	detected as an interesting entity	Innate biosynthesis	Acquisition from host	
Chemical mimesis[b]	Chemical mimicry or camouflage	Chemical mimicry	Chemical camouflage	Akino 2008[c] [14]
—	Chemical mimicry	No distinction		Bagnères and Lorenzi 2010[d] [34]
Chemical camouflage	Chemical mimicry	No distinction		Dettner and Liepert 1994 [15]
Chemical camouflage	Chemical mimicry	No distinction		Geiselhardt et al. 2007[e] [34]
—	Chemical mimicry	No distinction		Howard and Blomquist 2005 [32]
—	Chemical mimicry	No distinction		Keeling et al. 2004 [35]
—	Chemical mimicry	Chemical mimicry by biosynthesis	Chemical mimicry by camouflage	Lenoir et al. 2001 [11]
—	Chemical mimicry or camouflage	Chemical mimicry	Chemical camouflage	Nash and Boomsma 2008[c] [3]
—	Chemical mimicry	Not specified	—	Pierce et al. 2002 [36]
—	Chemical mimicry	Not specified	Chemical mimicry	Singer 1998[f] [37]
Chemical crypsis[g]	Chemical mimicry	No distinction		Stowe 1988 [31]
—	Chemical mimicry	Not specified	Chemical camouflage[h]	Thomas et al. 2005[c] [8]

—: not considered in the article. No distinction: the term chemical mimicry was used irrespective of the origin of mimetic cues. [a]According to the first two columns in Table 1. [b]Defined as being invisible through background matching. [c]Authors follow the definition of Howard et al. [38]. [d]Authors use the term mimicry irrespective of the origin of mimetic compounds but point out that different definitions exist depending on their origin. [e]Authors follow Dettner and Liepert [15]. [f]The term camouflage was used once to describe invading predators that biosynthesize CHCs of social insects. [g]Defined as resemblance of the background or of an entity in the background. [h]Inconsistent to the definitions of Dettner and Liepert [15].

TABLE 3: Proposed terminology for chemical adaptive resemblances. Chemical cues of a mimic can either be "*acquired*" from the environment (including the host), or they can be "*innate*", that is, biosynthesized. In all cases of chemical adaptive resemblance, the operator is deceived by the mimic so that the mimic benefits.

Suggested term	By an operator, the mimic is...
Chemical crypsis	... not detected as a discrete entity due to the expression of cues that blend with the environment (causing no reaction in the operator).
Chemical masquerade	... detected but misidentified as an uninteresting entity (causing no reaction in the operator).
Chemical mimicry	... detected as an entity of interest (causing a reaction in the operator).

empirically separate cases of masquerade and crypsis [28], but this has yet to be done in a nonvisual context. Table 3 gives an overview on our proposed terminology for chemical adaptive resemblances. Please note that in our terminology it is only important whether and how mimics are perceived by an operator. Similarities in the chemical profiles of parasites and hosts may be important diagnostic tools, but they are not part of the definitions.

Finally, we want to stress the special case of organisms that suppress the expression of chemical cues which can potentially be detected by the operator. Following our aim of applying a consistent biological terminology, "*chemical hiding*" is the most appropriate definition. This definition includes two slightly different scenarios, the total absence of relevant cues and the presence of cues below the operator's perceptive threshold. In both cases chemical perception of the organism is impossible. A host's inability to detect any chemical cues of a parasite was also referred to as "chemical insignificance" [3]. However, the term chemical insignificance is unfortunately used ambiguously regarding the important point whether there are no detectable cues [3] or small yet detectable amounts of cues are present [39]. Clearly, it should be distinguished whether an operator is able to detect an organism or not. If resemblance cues are present and perceived (irrespective of the quantitative level), the phenomenon will fall per definition into one of the categories chemical crypsis, chemical masquerade, or chemical mimicry (Table 3). For example, if a callows' weak chemical signature was expressed by a parasite and adult host ants misidentified this parasite as a callow, we would follow Ruxton [17] by assigning this to chemical mimicry (since callows are certainly interesting entities). Empirical evidence for a chemical mimicry of callows could result in practice from a combination of chemical data (callow resemblance) and behavioral data (hosts treat parasite as callows). However, an exhaustive discussion about methods is beyond the scope of this conceptual paper. Consequently, the original definition of chemical insignificance as a "weak signal" [39] appears not applicable to parasites without the risk of confusing it with chemical mimicry. If chemical cues are below an operator's perceptive threshold, the definition of chemical hiding will apply. However, the term chemical insignificance may be used as a functional term describing the lack of chemical information in a certain context. For example, callows are chemically insignificant in terms of nestmate recognition due to a lack of chemical information in that context. Nevertheless, callows carry apparently sufficient information in the context of caste identity since workers show characteristic behaviors towards them; for example, they receive assistance during hatching and are transported to new nest sites in migratory ants.

The above discussion on chemical insignificance applies also to the phenomenon of chemical transparency. If no cues are expressed that are perceivable by the operator, the focal organism would show chemical hiding, regardless of the presence of any other compounds. In contrast, if perceivable cues are present, chemical crypsis, chemical masquerade, or chemical mimicry applies. In the described case of chemical transparency [41], the parasite is most likely recognized and misidentified as an interesting entity (e.g., as brood), since social parasites usually exploit the brood care behavior of their hosts.

Notably, a parasite may alternatively avoid chemical detection through behavioral mechanisms by "hiding" according to the definition in general biology (see above) rather than "chemical hiding." For example, if it avoids detection by staying in a cavity so that its chemical cues do not reach the operator, it is hiding. A parasite that performs "hiding" could potentially be detected if it was somehow confronted with the operator. In contrast, a parasite that shows "chemical hiding" cannot be detected by chemical senses of the operator at all.

5. Examples for the Use of Adaptive Resemblance Terms

In this section we want to discuss examples to clarify the use of terms regarding adaptive resemblances. The mimicking of CHC profiles of the host is widespread among ant parasites, and this is generally assumed to facilitate integration into the host colonies. Parasites are indeed frequently not recognized as alien species [11, 33]. This strategy of avoiding recognition as an alien species by expression of host CHCs could potentially be referred to as chemical crypsis (if the colony odor is regarded as the background) or as chemical masquerade (if a nestmate worker is regarded as an uninteresting entity). However, we argue that the strategy is best described by chemical mimicry for the following reasons. First, workers are certainly able to detect other workers, and hence parasites that mimic them are discrete entities, excluding the term chemical crypsis. Second, workers are certainly interesting entities to other workers because social actions are shared, such as grooming or trophallaxis. Consequently, a mimic that uses a worker as model resembles an entity of potential interest to ant workers, so that chemical mimicry rather than chemical masquerade applies.

It becomes more complicated when a parasite mimics the nest odor of its host. Lenoir et al. [42] demonstrated that the inner nest walls of the ant species *Lasius niger* are coated with the same CHCs as those that occur on the cuticle of workers. However, the CHCs on the walls occurred in different proportions and showed no colony specificity. If a mimic resembles such a chemical profile, chemical crypsis will be the most appropriate term, because the mimic represents no discrete entity and rather blends with the uniform nest odor. To our knowledge, no clear evidence exists for this case.

Another example is worth highlighting in this context which was already pointed out by Ruxton [17]. The CHCs of *Biston robustum* caterpillars resemble the surface chemicals of twigs from its host plant [43]. *Formica japonica* and *Lasius japonicus* workers do not recognize the caterpillars on their native host plant, but when caterpillars were transferred to a different plant, the ants noticed and attacked them. In this case it depends on the operator's perception whether the example should be considered as chemical crypsis or chemical masquerade. If the ants did not detect a twig (and hence a caterpillar) as a discrete entity, but as background, chemical crypsis would apply. If the ants detected the caterpillar as a discrete but uninteresting entity, for example, as a twig, then chemical masquerade would apply. As Ruxton [17] emphasized, twigs are of huge dimension compared to the size of ants. Hence, it is more likely that ants do not detect caterpillars as discrete (uninteresting) entities, but rather perceive them as (uninteresting) background. Accordingly, chemical crypsis appears to be the most appropriate term for this example.

These examples may demonstrate that it can be rather difficult to assign appropriate terms to particular adaptive resemblance systems. Nevertheless, the definitions we proposed are generally straightforward, and they can be applied unambiguously if the necessary information about a system is available. We hope that this paper contributes to a careful and consistent use of adaptive resemblance terminology in chemical ecology.

Acknowledgments

The authors thank the behavioral ecology group at the LMU Munich and Graeme D. Ruxton for valuable comments. They are grateful to the editor Alain Lenoir and two anonymous reviewers for their effort to improve this paper. Thanks to Tomer Czaczkes for checking the orthography. The authors are grateful for financial support from the DFG (Deutsche Forschungsgemeinschaft, Project WI 2646/3).

References

[1] E. O. Wilson, "Success and dominance in ecosystems: the case of the social insects," in *Excellence in Ecology*, Ecology Institute, Oldendorf/Luhe, Germany, 1990.

[2] P. S. Ward, "Ants," *Current Biology*, vol. 16, no. 5, pp. R152–R155, 2006.

[3] A. Nash and J. J. Boomsma, "Communication between hosts and social parasites," in *Sociobiology of Communication an Interdisciplinary Approach*, P. d'Ettorre and D. P. Hughes, Eds., Oxford University Press, New York, NY, USA, 2008.

[4] P. Schmid-Hempel, *Parasites in Social Insects*, Princeton University Press, Princeton, NJ, USA, 1998.

[5] B. Hölldobler and E. O. Wilson, *The Ants*, Harvard University Press, Cambridge, Mass, USA, 1990.

[6] J. J. Boomsma, P. Schmid-Hempel, and W. O. H. Hughes, "Life histories and parasite pressure across the major groups of social insects," in *Insect Evolutionary Ecology: Proceedings of the Royal Entomological Society*, M. D. E. Fellowes, G. J. Holloway, and J. Rolff, Eds., pp. 139–176, CABI Publishing, Wallingford, Conn, USA, 2005.

[7] E. Wasmann, "Die Ameisen-und Termitengäste von Brasilien. I. Teil. Mit einem Anhange von Dr. August Forel.," *Verhandlungen der kaiserlich-königlichen zoologisch-botanischen Gesellschaft in Wien*, vol. 45, pp. 137–178, 1895.

[8] J. A. Thomas, K. Schönrogge, and G. W. Elmes, "Specializations and host associations of social parasites of ants," in *Insect Evolutionary Ecology Proceedings of the Royal Entomological Society*, M. D. E. Fellowes, G. J. Holloway, and J. Rolff, Eds., CABI Publishing, Sheffield, UK, 2005.

[9] R. Dawkins and J. R. Krebs, "Arms races between and within species," *Proceedings of the Royal Society of London B*, vol. 205, no. 1161, pp. 489–511, 1979.

[10] J. S. van Zweden and P. d'Ettorre, "Nestmate recognition in social insects and the role of hydrocarbons," in *Insect Hydrocarbons Biology, Biochemistry and Chemical Ecology*, G. J. Blomquist and A.-G. Bagnères, Eds., Cambridge University Press, New York, NY, USA, 2010.

[11] A. Lenoir, P. D'Ettorre, and C. Errard, "Chemical ecology and social parasitism in ants," *Annual Review of Entomology*, vol. 46, pp. 573–599, 2001.

[12] K. Dinter, W. Paarmann, K. Peschke, and E. Arndt, "Ecological, behavioural and chemical adaptations to ant predation in species of Thermophilum and Graphipterus (Coleoptera: Carabidae) in the Sahara desert," *Journal of Arid Environments*, vol. 50, no. 2, pp. 267–286, 2002.

[13] F. Barbero, J. A. Thomas, S. Bonelli, E. Balletto, and K. Schönrogge, "Queen ants make distinctive sounds that are mimicked by a butterfly social parasite," *Science*, vol. 323, no. 5915, pp. 782–785, 2009.

[14] T. Akino, "Chemical strategies to deal with ants: a review of mimicry, camouflage, propaganda, and phytomimesis by ants (Hymenoptera:Formicidae) and other arthropods," *Myrmecological News*, vol. 11, pp. 173–181, 2008.

[15] K. Dettner and C. Liepert, "Chemical mimicry and camouflage," *Annual Review of Entomology*, vol. 39, pp. 129–154, 1994.

[16] S. Steiger, T. Schmitt, and H. M. Schaefer, "The origin and dynamic evolution of chemical information transfer," *Proceedings of the Royal Society B*, vol. 278, no. 1708, pp. 970–979, 2011.

[17] G. D. Ruxton, "Non-visual crypsis: a review of the empirical evidence for camouflage to senses other than vision," *Philosophical Transactions of the Royal Society B*, vol. 364, no. 1516, pp. 549–557, 2009.

[18] A. Starrett, "Adaptive resemblance: a unifying concept for mimicry and crypsis," *Biological Journal of the Linnean Society*, vol. 48, no. 4, pp. 299–317, 1993.

[19] W. Wickler, *Mimikry. Nachahmung und Täuschung in der Natur*, Kindlers Universitäts Bibliothek, München, Germany, 1968.

[20] R. I. Vane-Wright, "On the definition of mimicry," *Biological Journal of the Linnean Society*, vol. 13, no. 1, pp. 1–6, 1980.

[21] J. A. Endler, "An overview of the relationships between mimicry and crypsis," *Biological Journal of the Linnean Society*, vol. 16, no. 1, pp. 25–31, 1981.

[22] J. A. Endler, "Frequency-dependent predation, crypsis and aposematic coloration," *Philosophical Transactions of the Royal Society of London B*, vol. 319, no. 1196, pp. 505–523, 1988.

[23] G. Pasteur, "A classificatory review of mimicry systems," *Annual Review of Ecology and Systematics*, vol. 13, pp. 169–199, 1982.
[24] M. H. Robinson, "A stick is a stick and not worth eating: on the definition of mimicry," *Biological Journal of the Linnean Society*, vol. 16, no. 1, pp. 15–20, 1981.
[25] G. D. Ruxton, T. N. Sherratt, and M. P. Speed, *Avoiding Attack: The Evolutionary Ecology of Crypsis, Warning Signals and Mimicry*, Oxford University Press, New York, NY, USA, 2004.
[26] M. Stevens and S. Merilaita, "Animal camouflage: Current issues and new perspectives," *Philosophical Transactions of the Royal Society B*, vol. 364, no. 1516, pp. 423–427, 2009.
[27] R. I. Vane-Wright, "A unified classification of mimetic resemblances," *Biological Journal of the Linnean Society*, vol. 8, no. 1, pp. 25–56, 1976.
[28] J. Skelhorn, H. M. Rowland, and G. D. Ruxton, "The evolution and ecology of masquerade: review article," *Biological Journal of the Linnean Society*, vol. 99, no. 1, pp. 1–8, 2010.
[29] H. W. Bates, "Contributions to an insect fauna of the Amazon valley (Lepidoptera: Heliconidae)," *Transactions of the Linnean Society of London*, vol. 23, no. 3, pp. 495–566, 1862.
[30] M. R. E. Symonds and M. A. Elgar, "The evolution of pheromone diversity," *Trends in Ecology and Evolution*, vol. 23, no. 4, pp. 220–228, 2008.
[31] M. K. Stowe, "Chemical mimicry," in *The Chemical Mediation of Coevolution*, K. C. Spencer, Ed., Academic Press, New York, NY, USA, 1988.
[32] R. W. Howard and G. J. Blomquist, "Ecological, behavioral, and biochemical aspects of insect hydrocarbons," *Annual Review of Entomology*, vol. 50, pp. 371–393, 2005.
[33] A.-G. Bagnères and M. Lorenzi, "Chemical deception/mimicry using cuticular hydrocarbons," in *Insect hydrocarbons Biology, Biochemistry and Chemical Ecology*, G. J. Blomquist and A.-G. Bagnères, Eds., Cambridge University Press, New York, NY, USA, 2010.
[34] S. F. Geiselhardt, K. Peschke, and P. Nagel, "A review of myrmecophily in ant nest beetles (Coleoptera: Carabidae: Paussinae): linking early observations with recent findings," *Naturwissenschaften*, vol. 94, no. 11, pp. 871–894, 2007.
[35] C. I. Keeling, E. Plettner, and K. N. Slessor, "Hymenopteran semiochemicals," *Topics in Current Chemistry*, vol. 239, pp. 133–177, 2004.
[36] N. E. Pierce, M. F. Braby, A. Heath et al., "The ecology and evolution of ant association in the Lycaenidae (Lepidoptera)," *Annual Review of Entomology*, vol. 47, pp. 733–771, 2002.
[37] T. L. Singer, "Roles of hydrocarbons in the recognition systems of insects," *American Zoologist*, vol. 38, no. 2, pp. 394–405, 1998.
[38] R. W. Howard, R. D. Akre, and W. B. Garnett, "Chemical mimicry in an obligate predator of carpenter ants (Hymenoptera: Formicidae)," *Annals of the Entomological Society of America*, vol. 83, pp. 607–616, 1990.
[39] A. Lenoir, D. Fresneau, C. Errard, and A. Hefetz, "Individuality and colonial identity in ants: the emergence of the social representation concept," in *Information Processing in Social Insects*, C. Detrain, J. L. Deneubourg, and J. Pasteels, Eds., Birkhäuser, Basel, Switzerland, 1999.
[40] R. M. Kilner and N. E. Langmore, "Cuckoos versus hosts in insects and birds: adaptations, counter-adaptations and outcomes," *Biological Reviews*, vol. 86, no. 4, pp. 836–852, 2011.
[41] S. J. Martin, J. I. Takahashi, M. Ono, and F. P. Drijfhout, "Is the social parasite Vespa dybowskii using chemical transparency to get her eggs accepted?" *Journal of Insect Physiology*, vol. 54, no. 4, pp. 700–707, 2008.
[42] A. Lenoir, S. Depickère, S. Devers, J. P. Christidès, and C. Detrain, "Hydrocarbons in the ant Lasius niger: from the cuticle to the nest and home range marking.," *Journal of Chemical Ecology*, vol. 35, no. 8, pp. 913–921, 2009.
[43] T. Akino, K. I. Nakamura, and S. Wakamura, "Diet-induced chemical phytomimesis by twig-like caterpillars of Biston robustum Butler (Lepidoptera: Geometridae)," *Chemoecology*, vol. 14, no. 3-4, pp. 165–174, 2004.

Summarized results

Chapter 1: Differential host defence against multiple parasites in ants

The results of this study indicate that the hosts' defence and the impact of parasites are connected in a multi-parasite system in that parasites imposing high costs are more likely to be fended off by the host. Staphylinid beetle species that preyed on ant larvae were often attacked, resulting in low levels of integration, i.e. they stayed outside of the nest and avoided contact to host workers. In contrast, staphylinid beetles that did not prey on ant larvae but instead fed solely on host prey items were seldom attacked and achieved a high level of social integration, i.e. they stayed inside the nest and interacted frequently with host workers.

Chapter 2: Acquisition of chemical recognition cues facilitates integration into ant societies

By analyzing the transfer of a labelled hydrocarbon from host ants to parasites, it was demonstrated that the kleptoparasitic silverfish *Malayatelura ponerophila* acquires CHCs from its host. The concentration of each CHC decreased in isolated silverfish, indicating that no additional biosynthesis of mimetic compounds occurs. Furthermore, the study revealed that the accuracy of chemical mimicry is crucial for this parasite to avoid host aggression and gain social acceptance. Silverfish with experimentally lowered chemical host similarity were attacked more frequently than non-isolated (unmanipulated) individuals.

Chapter 3: The social integration of a myrmecophilous spider does not depend exclusively on chemical mimicry

The spider *Gamasomorpha maschwitzi* acquired a chemical label from its host, strongly indicating that it is able to acquire mimetic CHCs similarly to the silverfish *M. ponerophila*. Additional biosynthesis of mimetic CHCs seems unlikely, since the concentration of each CHC decreased in isolated individuals. However, clear differences in the social integration mechanisms of the spider and the silverfish were found. Contrary to my expectations, the spiders remained socially integrated in the host colonies despite experimentally reduced chemical host resemblance. Although the spiders apparently do not depend as much as the silverfish on a high accuracy of chemical host resemblance, the interactions between spiders and ants suggest that they nevertheless benefit to some degree from chemical mimicry.

Chapter 4: On the use of adaptive resemblance terms in chemical ecology

Since adaptive resemblance terms are used inconsistently within chemical ecology literature and with respect to the usage in general biology, the following consistent terminology was suggested: "Chemical crypsis" occurs when the operator does not detect the mimic as a discrete entity (background matching). "Chemical masquerade" occurs when the operator detects the mimic but misidentifies it as an uninteresting entity, as opposed to "chemical mimicry" in which an organism is detected as an interesting entity by the operator. The additional terms "acquired" and "innate" may be used to specify the origins of mimetic cues.

General discussion

"Obviously, then, some symbionts possess the key that unlocks the fortress door. Using covert means, they have gained entrance to a notoriously well-defended colony. Once through the portal, still others, consummate thespians, play an adaptive charade, pretending to be what they are not: members of the colony."
W.H. Gotwald, Jr. (1995)

The spider and the silverfish seek contact with host workers to acquire mimetic CHCs.

The results of my dissertation shed light on important coevolutionary interactions between ants and their parasites. In particular, the underlying mechanisms accounting for the different levels of social integration among the studied parasites have been elucidated (see summarized results). Table II shows the initial proposed hypotheses together with an assessment of whether they were supported or not. In the following sections I will discuss the ultimate and proximate mechanisms that facilitate social integration.

Table II. Proposed hypotheses for the different research approaches and validation.

Research approach	Research topic	Hypotheses	Validation
Ultimate mechanisms (Chapter 1)	Interdependency of parasite impact and host defence	More costly myrmecophiles are attacked more frequently. Accordingly, they achieve lower integration levels.	Yes.
Proximate mechanisms (Chapter 2 and 3)	Origin of mimetic compounds	The two myrmecophiles acquire mimetic compounds from their host.	Yes.
	Accuracy in chemical mimicry facilitates integration	Myrmecophiles showing reduced chemical host resemblance are attacked more frequently and thus achieve lower levels of integration.	Yes, regarding the silverfish. No, regarding the spider.

Ultimate mechanisms: Why are some myrmecophiles integrated and others are not?

"Is there a hierarchy of [host defense] *behaviors that we expect to see, depending on the magnitude of the parasitic threat and the cost of the response?"*
Moore (2002)

Different levels of social integration existed among the studied beetle species (chapter 1), which apparently depended on the myrmecophiles' fitness impact. Under the pressure of a diverse parasite community, *Leptogenys* ants attacked preferentially costly (predatory) beetle species while less costly (non-predatory) species were treated peacefully, consequently achieving higher levels of social integration. I will discuss the evolutionary consequences of multi-parasite situations from the perspective of the host as well as from that of the parasites.

From the perspective of the host, infections with parasites are by definition associated with fitness costs (see e.g. Hughes et al. 2008). As a counter-adaptation against parasites, hosts have evolved complex behavioural, morphological and physiological defence strategies to prevent parasite encounters and/or to defend and consequently control the parasite load once encounters have taken place (Hart 1990; Sheldon and Verhulst 1996; Combes 2005; Cremer et al. 2007; Abbas et al. 2011). However, the development and maintenance of defence mechanisms are expected to be costly (Lochmiller and Deerenberg 2000; Cremer et al. 2007). To optimize the energetic costs one can expect a trade-off between the energy invested in defences and the resulting benefits of reduced parasite impact. Accordingly, hosts should be driven towards highly efficient defence strategies against parasites. In multi-parasitized hosts various defence strategies are conceivable to achieve this goal.

First of all, an efficient way for any host to deal with parasites is to fine-tune the energy invested in defence according to the current parasite load. One would expect that under a high parasite load the energy invested in defence would be increased. Such an adjustable or inducible defence has been demonstrated in various kinds of organisms and in different defence systems (e.g., immune system in vertebrates: Abbas et al. 2011; social immune system in social insects: Cremer et al. 2007; host aggressiveness against parasites in ants: Pamminger et al. 2011). Whether the defence behaviour of *L. distinguenda* is also adjustable and whether it depends on the actual parasite load (e.g., elevated aggression under a higher parasite load) is unknown and worth studying in future projects.

Another option for coping with multiple parasites is to direct the same type of defence in a non-specific manner against many or all parasitic species to lower total costs, and to increase fitness accordingly. The vertebrates' innate immune system is a well studied example of such a non-specific defence as it recognizes and attacks any foreign body (antigen) in a non-specific manner (Abbas et al. 2011). Another well known non-specific defence is grooming behaviour, including self-grooming and grooming of group members which occur, for example, in primates, birds and social insects (Schmid-Hempel 1998; Clayton et al. 1999; Nunn and Altizier 2006). Although not shown in our experiments, the frequent colony migrations of the host ant *L. distinguenda* to new nest sites (Steghaus-Kovac 1994) might be a non-specific defence mechanism against parasites as well. Even though most myrmecophiles evolved elaborate mechanisms to participate in ant migrations (Witte et al. 2008), a reduction in parasite number due to the frequent migrations cannot be ruled out and is thus worth examination in future projects.

Alternatively, hosts can direct their defence specifically at the most costly parasites in order to effectively reduce the total parasite costs. An adjustable host defence depending on the impact inflicted by an opponent has been shown recently in a study on *Temnothorax* ants (Scharf et al. 2011). *Temnothorax* workers attacked their highly costly social parasite *P. americanus* more often than less damaging competitors. The study on myrmecophilous staphylinid beetles (chapter 1) provided the first evidence that an adjustment of host defence dependant on the parasites' impact can occur in a multi-parasite situation. Regarding the similar body sizes and abundances of the five studied beetle species, beetles preying on host brood should impose significantly higher costs to the host than kleptoparasitic beetles, which feed solely on the ants' prey. Since the fitness of army ant colonies strongly depends on the colony growth (Gotwald 1995), a loss of brood through predation can be considered to impose substantial costs. Accordingly, selection pressure on the host to evolve counter-defences should be stronger against brood-killing beetles than against kleptoparasitic beetles. Indeed, potentially costly beetles were effectively excluded from the nest interior of host colonies, which houses the colonies' brood. Since less aggressive host colonies would suffer the costs of brood predation, colonies attacking and fending off predatory beetles should have selective advantages, and thus the specific host defence against costly beetles should be maintained within the host population. Evolving towards lower virulence might have been highly beneficial for staphylinid beetles (predation is a plesiomorphic trait in staphylinids, see chapter 1) since selection pressure on the host to evolve counter-defences against them should be weaker. Under these circumstances myrmecophiles might be able to achieve higher levels of integration allowing them to enter the nest interior of their host. Living in this homeostatic, protected environment would provide many benefits such as protection from own parasites, competitors or predators or the reliable provision of high quality food (Hölldobler and Wilson 1990; Boomsma et al. 2005).

Ultimate mechanisms: Conclusion

A tendency of social insect parasites to evolve less virulence due to an effective host defence and to additional factors associated with the biology of social insects was already proposed by Hughes et al. (2008). Although important aspects of parasitology remain unknown in the studied army ant-staphylinid beetle system, the results provide the first evidence that selection towards less virulent parasites may indeed occur in a multi-parasite situation due to a differentiated host defence that specifically target costly parasites. Such a differentiated defence can also explain the different levels of social integration found among myrmecophiles.

Ultimate mechanisms: Future directions

An interesting approach for future research would be to test whether the interdependency of parasites' impact and host defence also holds for other taxonomic groups. Preliminary observations on other myrmecophiles of *L. distinguenda* (e.g., spider, silverfish, phorid fly, and mite) indeed indicated that the hosts' defence is stronger against more costly parasites irrespective of the parasites' taxon. However, more detailed studies are necessary to verify these observations.

Proximate mechanisms: Why are some myrmecophiles integrated and others are not?

"One might almost imagine they [two specific myrmecophiles] *had the cap of invisibility."*
Lubbock (1891)

A high degree of accuracy in chemical host resemblance was shown to be beneficial for the two studied myrmecophiles, the silverfish *M. ponerophila* and the spider *G. maschwitzi* (Chapter 2 and 3). However, only the silverfish relied on high accuracy in chemical host resemblance to avoid ant attacks and consequently achieve social integration. By evaluating the transfer of a chemical label from host ants to parasites it was additionally demonstrated that both myrmecophiles are able to acquire mimetic CHCs from their host. In the following I will discuss the evolutionary consequences of differing origins of mimetic compounds and the importance of accuracy in chemical mimicry.

Origin of mimetic compounds

Acquisition from the environment as well as biosynthesis of mimetic compounds is expected to be associated with initial costs for mimics (Ruxton et al. 2004). Regarding myrmecophiles, an acquisition of host CHCs via physical contact can be risky because ants generally attack any foreign organism trespassing on their nest (Lubbock 1891; Hölldobler and Wilson 1990). Myrmecophiles relying on 'acquired chemical mimicry' (sensu chapter 4) as integration mechanism face two main problems. First, they have to cope with their host while invading a colony. At this stage, they have no prior host contact and therefore no chance to acquire the current colony odour. As a consequence, recognition as an intruder by ant workers is likely. To avoid being attacked or killed while invading new host colonies myrmecophiles have evolved elaborate strategies which can also be expected to be costly (see chapter 2). Once an invasion is successful, myrmecophiles have to regularly replenish their mimetic profile. As shown for the two studied myrmecophiles, this requires staying in close contact with their host (chapter 2 and 3). Specific behaviours facilitating the acquisition of host CHCs such as intense rubbing on host workers have also been described for many other myrmecophiles from different taxa (Akre and Rettenmeyer 1966; Hölldobler and Wilson 1990; Akino and Yamaoka 1998; Witte et al. 2009). Such close host contact is associated with the risk of being recognized and attacked. Furthermore, it requires myrmecophiles to constantly move to avoid losing contact, which is surely accompanied with physiological costs (see also the supplemental videos).

Biosynthesis of mimetic cues, on the other hand, is associated with costs as well, among others with the metabolic energy of CHC synthesis (see Blomquist and Bagnères 2010). Furthermore, a biosynthesis of complex host recognition signatures by distantly related myrmecophiles is a rather unlikely event due to the following reasons. Evolving biosynthetic pathways for the production of the essential host recognition cues requires several genes to mutate appropriately. The likelihood to evolve such mutations strongly depends on the time span of coevolution and on the relatedness of host and parasite. However, even if a distantly related myrmecophile had evolved the biosynthetic pathways to produce the relevant host recognition cues, it is unlikely that it would be able to express the compounds in the correct relative proportions. The so called 'gestalt colony odour' (Crozier and Dix 1979) of ants is dynamic and can change spontaneously according to genetical changes in the colony (e.g., a new queen) and/or shifts in the environment (e.g., in the diet or the nest materials; reviewed in van Zweden and d'Ettorre 2010). Organisms that do not match this flexible but specific colony odour are generally attacked by ants, at least inside their nest (Hölldobler and Wilson 1990; Lenoir et al. 2001). Myrmecophiles mimicking the current colony odour were, for example, seldom attacked in their resident colony. However, if transferred to another conspecific host colony, they were instantly attacked and killed demonstrating the importance of specificity in the colony odour (Akino and Yamaoka 1998; Witte et al. 2009). Even if key regulatory enzymes were involved, biosynthesis of mimetic compounds would unlikely be

General discussion

adjustable to the current colony odour. It is worth highlighting the special case of slave-making ants in this context. In contrast to most other myrmecophiles, slave-making ants are often closely related to their host species (a phenomenon called 'Emery's rule'; see Hölldobler and Wilson 1990), resulting in similar communication systems of host and parasite (Buschinger 2009). This includes similarities in nestmate recognition cues, which render biosynthesis of host recognition cues more likely. Although similarities of CHCs often exist between host and slavemaker (D'Ettorre et al. 2002; Bauer et al. 2010), the slavemaker still has to deal with the specific but dynamic gestalt odour of their host colonies. An inability to match the current gestalt odour via biosynthesis of mimetic CHCs might be the reason why many slave-making ants do not rely solely on 'innate chemical mimicry' (sensu chapter 4) but on otherwise evolved elaborate strategies to invade host colonies (reviewed in Buschinger 2009). In the small number of studies describing the biosynthesis of mimetic cues among myrmecophiles (reviewed in Akino 2008), only few key stimuli such as brood or male pheromones are mimicked to achieve adoption or integration (Howard et al. 1990b; Akino et al. 1999; Schönrogge et al. 2004; Nash et al. 2008). Since brood recognition cues in ants are generally not colony-specific (sometimes not even species-specific; reviewed in Hölldobler and Wilson 1990), the biosynthesis of a few key stimuli can be expected to be an efficient way for myrmecophiles to invade different host colonies. Indeed, some of these brood-mimicking myrmecophiles have been shown to be picked up in the environment and carried into the nest by ant workers from different host colonies (Akino and Yamaoka 1998; Nash et al. 2008). Myrmecophiles identified as ant brood by workers may receive additional benefits, e.g. a benign treatment, food provision by workers as well as the chance to be carried if the colony migrates to a new nest site.

Interestingly, some social insect parasites apply a combination of both strategies (acquired and innate mimicry) to attain mimetic CHCs (reviewed in Bagnères and Lorenzi 2010). The myrmecophilous blue butterfly *Phengaris* (formerly *Maculinea*) *rebeli* (Lepidoptera:

Lycaenidae) is among the most well understood examples. The caterpillars synthesize a chemical profile that is attractive to ants which pick them up and transport the parasite into their nest (Nash et al. 2008). The caterpillars' pre-adoption profile consists of several compounds that are likely key stimuli in the hosts' recognition of brood or of young workers (Akino et al. 1999; Elmes et al. 2002; Nash et al. 2008). Once adopted, the caterpillar additionally acquires colony-specific compounds from their host (Akino et al. 1999; Schönrogge et al. 2004). Whether this myrmecophile gains additional benefits by acquiring further host CHCs is so far unknown. The acquisition of further mimetic cues probably makes identification as an intruder more difficult and/or increases the frequency of benevolent behaviours. In summary, the biosynthesis of few non-colony-specific key stimuli of the hosts' recognition system (e.g., brood or male cues) seems to be an elaborate chemical strategy among myrmecophiles to achieve adoption by ants. Once settled in a host colony, the acquisition of host CHCs appears to be another sophisticated integration mechanism for myrmecophiles.

Costs associated with mimicry in general must be compensated by benefits (Ruxton et al. 2004) and so must the costs associated with chemical mimicry which I described above. However, the benefits of chemical mimicry have rarely been demonstrated (see general introduction). For the first time, we demonstrated that chemical mimicry is indeed advantageous for two myrmecophilous species if the accuracy in chemical resemblance is high.

The role of accuracy in chemical mimicry

Mimicry systems (excluding Müllerian mimicry) can be considered as coevolutionary arms races in which the frequencies of model and mimic traits are continually changing over time (Ruxton et al. 2004). If the presence of mimics imposes a strong negative impact on the model organism, the models are expected to evolve new traits which are difficult to resemble (Ruxton et al. 2004). As counter-adaptations, mimics evolve towards improvement in mimicry accuracy, resulting in a dynamic equilibrium of reciprocal changes (Ruxton et al. 2004). Benefits for mimics that depend on the accuracy of resemblance have been demonstrated comprehensively in auditory and visual sensory systems (Mappes and Alatalo 1997; Ruxton et al. 2004; Coleman et al. 2007). In contrast to these mimetic systems, studies on the accuracy of chemical adaptive resemblance have rarely been carried out, although chemical communication is the most widespread form of communication among organisms (Stowe 1988; Symonds and Elgar 2008; Steiger et al. 2011). The myrmecophilous butterfly *Phengaris alcon* (Lepidotera: Lycanidae) is one of the few examples, for which the accuracy of chemical resemblance has been shown to be beneficial for the mimic in that closer chemical host resemblance resulted in quicker adoption (Nash et al. 2008). Since mortality during the caterpillars' adoption stage is the main key factor in the life-history of these butterflies (Elmes et al. 2002), a quick and efficient adoption is expected to be highly beneficial for them (Nash et al. 2008). Our study on the myrmecophilous silverfish *M. ponerophila* demonstrated for the first time that the social integration of myrmecophiles can strongly depend on the accuracy of chemical host resemblance (Fig. IV). While silverfish individuals with reduced chemical host resemblance were frequently attacked by ants and sometimes even killed, individuals showing a higher accuracy in chemical host resemblance were seldom attacked, resulting in high levels of social integration. Thus, silverfish gain the benefits of social life, e.g. the protection by host ants from predators and the reliable provision of high quality food.

Although not as obvious as for the silverfish, a higher accuracy in chemical mimicry was also beneficial for the spider since unmanipulated spiders (with higher host resemblance) are less often inspected by ant workers (chapter 3). Apparently, the spider depended less on chemical cues to avoid ant attacks compared to the silverfish, suggesting that additional factors may play important roles for their high level of social integration. Indeed, many myrmecophiles additionally show other integration mechanisms such as behavioural, acoustical and morphological adaptations (Hölldobler and Wilson 1990; Gotwald 1995; Barbero et al. 2009). The two studied myrmecophiles showed conspicuous differences in behaviour and morphology which likely explain their different dependency on chemical deception (see discussion in chapter 3).

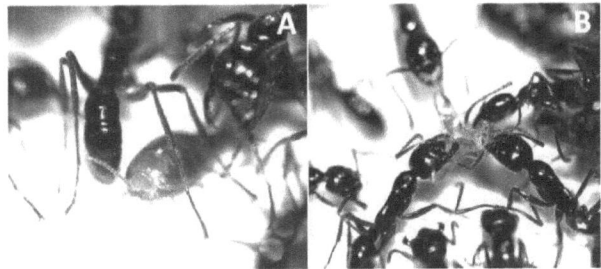

Figure IV. Physical contact with host workers was necessary for the silverfish to acquire mimetic CHCs, which are used by ants as recognition cues. (A) Non-isolated (unmanipulated) silverfish mimicked the chemical profiles of their host well enough to achieve high levels of social integration. (B) Individuals with experimentally reduced chemical host resemblance were recognised, attacked, and sometimes killed by the host ants.

Proximate mechanisms: Conclusion

Although elaborate behavioural and morphological adaptations may be required, I argued that the acquisition of mimetic compounds from the hosts is an evolutionarily parsimonious mechanism for myrmecophiles to achieve and maintain social integration. Furthermore, accuracy in chemical resemblance can be crucial for myrmecophilous intruders to achieve social integration into ant societies. However, the degree of dependency on chemical mimicry as an integration mechanism can differ considerably between myrmecophilous species.

Proximate mechanisms: Future directions

Future research projects dealing with the integration level of myrmecophiles best consider and link chemical, acoustical, tactile, morphological, as well as behavioural traits to potentially uncover which features are important (and thus likely adaptations) for a myrmecophilous species to cope with its host. One can predict, for example, that a myrmecophile with a protective morphology will depend less on avoidance behaviour or a high accuracy of chemical resemblance compared to a morphologically unprotected myrmecophile.

Another direction for future research would be a broad comparison between myrmecophiles and their closest non-myrmecophilous relatives. Such an approach would offer the possibility to investigate whether certain traits are likely adaptations to a myrmecophilous lifestyle or not, and in addition it would allow investigating whether specific features occur predominantly in certain taxa. For example, the convergent evolution of the limuloid (drop-shaped) body form in unrelated myrmecophiles provides strong evidence of its adaptive importance, but it could equally be an adaptation to the life in small cavities or an ancestral trait.

General outlook

This study on army ants and their parasites has demonstrated that these associations are well-suited to study host-parasite interactions in multi-parasite systems. They provide ample opportunities to test several predictions regarding the social integration mechanisms of ant parasites. For parasites that impose comparably low costs on their hosts, the evidence suggests that it is easier to evolve appropriate traits that allow them to achieve high levels of social integration and thus to gain the benefits of social life. Acquired or innate chemical mimicry by ant parasites, for example, has been demonstrated to be an elaborate adaptation that facilitates social integration. On the other hand, the evidence suggests that it is more difficult for comparably costly ant parasites to achieve social integration in multi-parasite systems, since they face stronger host defence. These parasites might evolve proximate mechanisms that allow them to cope but not to peacefully interact with their host during encounters (e.g., protective morphology or defensive glandular segregations). Alternatively, costly ant parasites could evolve towards less virulence. As a consequence, the host defence towards them is expected to decrease and thus social integration can more easily evolve. Since social integration into ant societies is expected to be highly beneficial for ant parasites, I expect that parasites that share a long coevolutionary history with their army ant hosts will likely show reduced virulence and elaborate mechanisms to gain social integration.

References

Abbas AK, Lichtmann AH, and Pillai S (2011). **Cellular and molecular immunology.** 7th edition edn. Elsevier Saunders, Philadelphia, USA

Akino T (2008). Chemical strategies to deal with ants: a review of mimicry, camouflage, propaganda, and phytomimesis by ants (Hymenoptera:Formicidae) and other arthropods. *Myrmecological News* 11:173-181

Akino T, Knapp JJ, Thomas JA, and Elmes GW (1999). Chemical mimicry and host specificity in the butterfly *Maculinea rebeli*, a social parasite of *Myrmica* ant colonies. *Proceedings of the Royal Society London Series B* 266:1419-1426

Akino T, and Yamaoka R (1998). Chemical mimicry in the root aphid parasitoid *Paralipsis eikoae* Yasumatsu (Hymenoptera: Aphidiidae) of the aphid-attending ant *Lasius sakagamii* Yamauchi & Hayashida (Hymenoptera: Formicidae). *Chemoecology* 8:153-161

Akre RD, and Rettenmeyer CW (1966). Behavior of Staphylinidae associated with army ants (Formicidae:Ecitoninae). *Journal of Kansas Entomological Society* 39:745-782

Allan RA, Capon RJ, Brown WV, and Elgar MA (2002). Mimicry of host cuticular hydrocarbons by salticid spider *Cosmophasis bitaeniata* that preys on larvae of tree ants *Oecophylla smaragdina*. *Journal of Chemical Ecolology* 28:835-848

Bagnères A-G, and Lorenzi M (2010). Chemical deception/mimicry using cuticular hydrocarbons. In: Blomquist GJ, Bagnères A-G (eds) **Insect hydrocarbons: Biology, Biochemistry and Chemical ecology.** Cambridge University Press, New York, USA

Barbero F, Thomas JA, Bonelli S, Balletto E, and Schönrogge K (2009). Queen ants make distinctive sounds that are mimicked by a butterfly social parasite. *Science* 323:782-785

Bauer S, Boehm M, Witte V, and Foitzik S (2010). An ant social parasite in-between two chemical disparate host species. *Evolutionary Ecology* 24:317-332

Blomquist GJ, and Bagnères A-G (2010). **Insect Hydrocarbons: Biology, Biochemistry and Chemical Ecology.** Cambridge University Press, New Yok, USA

Boomsma JJ and Nash DR (2008). Communication between hosts and social parasites. In: d'Ettorre P and Hughes DP (eds). **Sociobiology of communication.** Oxford University Press, New York, USA

Boomsma JJ, Schmid-Hempel P, and Hughes WOH (2005). Life histories and parasite pressure across the major groups of social insects. In: Fellowes MDE, Holloway GJ, and Rolff J (eds). **Insect Evolutionary Ecology: Proceedings of the Royal Entomological Society's 22nd Symposium.** CABI, Wallingford, UK

Buschinger A (2009). Social parasitism among ants: a review (Hymenoptera: Formicidae). *Myrmecological News* 12:219-235

Clayton DH, Lee PL, Tompkins DM, and Brodie IE (1999). Reciprocal natural selection on host-parasite phenotypes. *American Naturalist* 154:261-270

Coleman SW, Patricelli GL, Coyle B, Siani J, and Borgia G (2007) Female preferences drive the evolution of mimetic accuracy in male sexual displays. *Biology Letters* 3:463-466

Combes C (2005). **The art of being a parasite.** The University of Chicago Press, Chicago, USA

Cremer S, Armitage SA, and Schmid-Hempel P (2007). Social immunity. *Current Biology* 17:693-702

Crozier RH, Dix MW (1979). Analysis of two genetic models for the innate components of colony odor in social hymenoptera. *Behavioral Ecology and Sociobiology* 4:217-224

D'Ettorre P, Mondy N, Lenoir A, and Errard C (2002). Blending in with the crowd: social parasites integrate into their host colonies using a flexible chemical signature. *Proceedings of the Royal Society London Series B* 269:1911-1918

Dawkins R, and Krebs JR (1979). Arms races between and within species. *Proceedings of the Royal Society London Series B* 205:489-511

de Meeus T, and Renaud F (2002). Parasites within the new phylogeny of eukaryotes. *Trends in Parasitology* 18:247-251

de Roode JC, Pansini R, Cheesman SJ, Helinski MEH, Huijben S, Wargo AR, Bell AS, Chan BHK, Walliker D, and Read AF (2005). Virulence and competitive ability in genetically diverse malaria infections. *Proceedings of the National Academy of Sciences of the United States of America* 102:7624-7628

Dimijian GG (2000). Evolving together: the biology of symbiosis, part 1. *BUMC Proceedings* 13:217–226

Elmes GW, Akino T, Thomas JA, Clarke RT, and Knapp JJ (2002). Interspecific differences in cuticular hydrocarbon profiles of *Myrmica* ants are sufficiently consistent to explain host specificity by *Maculinea* (large blue) butterflies. *Oecologia* 130:525-535

Elmes GW, Clarke RT, Thomas JA, and Hochberg ME (1996). Empirical tests of specific predictions made from a spatial model of the population dynamics of *Maculinea rebeli*, a parasitic butterfly of red ant colonies. *Acta Oecologica* 17:61-80

Goff LJ (1982). Symbiosis and Parasitism: Another Viewpoint. *BioScience* 32:255-256

Gösswald K (1955). Gäste der Ameisen. In: **Unsere Ameisen II.** KOSMOS, Stuttgart, GER

Gotwald WH, Jr. (1995). **Army ants: the biology of social predation.** Cornell University Press, Ithaca, New York, USA

Hart BL (1990). Behavioral adaptations to pathogens and parasites: five strategies. *Neuroscience & Biobehavioral Reviews* 14:273-94

Hölldobler B, and Wilson EO (1990). **The ants.** Harvard University Press, Cambridge, Massachusetts, USA

Howard RW, Akre RD, and Garnett WB (1990a). Chemical mimicry in an obligate predator of carpenter ants (Hymenoptera: Formicidae). *Annals of the Entomological Society of America* 83:607-616

Howard RW, Stanley-Samuelson DW, and Akre RD (1990b). Biosynthesis and chemical mimicry of cuticular hydrocarbons from the obligate predator, *Microdon albicomatus* Novak (Diptera: Syrphidae) and its ant prey, *Myrmica incompleta* Provancher (Hymenoptera: Formicidae). *Journal of the Kansas Entomological Society* 63:437-443

Hughes DP, Pierce NE, and Boomsma JJ (2008). Social insect symbionts: evolution in homeostatic fortresses. *Trends in Ecology & Evolution* 23:672-677

Kistner DH (1979). Social and evolutionary significance of social insect symbionts. In: Hermann HR (ed) **Social insects.** Academic press, New York, USA, pp 339-413

Kistner DH, Witte V, and Maschwitz U (2003). A new species of *Trachydonia* (Coleoptera : Staphylinidae, Aleocharinae) from Malaysia with some notes on its behavior as a guest of *Leptogenys* (Hymenoptera : Formicidae). *Sociobiology* 42:381-389

Kronauer DJC (2009). Recent advances in army ant biology (Hymenoptera: Formicidae). *Myrmecological News* 12:51-65

Laforsch C, and Tollrian R (2004). Inducible defenses in multipredator environments: Cyclomorphosis in *Daphnia cucullata*. *Ecology* 85:2302-2311

Lenoir A, D'Ettorre P, Errard C, Hefetz A (2001) Chemical ecology and social parasitism in ants. *Annual Review of Entomology* 46:573-599

Lochmiller RL, and Deerenberg C (2000). Trade-offs in evolutionary immunology: just what is the cost of immunity? *Oikos* 88:87-98

Lubbock J (1891). **Ants, bees, and wasps: A record of observations on the habits of the social hymenoptera.** Kegan Paul, Trench, Trübner, & Co. Ltd., London, UK

Mappes J, and Alatalo RV (1997). Batesian mimicry and signal accuracy. *Evolution* 51:2050-2053

Martens K, and Schon I (2000). Parasites, predators and the Red Queen. *Trends in Ecology & Evolution* 15:392-393

Maruyama M, von Beeren C, and Hashim R (2010a). Aleocharine rove beetles (Coleoptera, Staphylinidae) associated with *Leptogenys* Roger, 1861 (Hymenoptera, Formicidae) I. Review of three genera associated with *L. distinguenda* (Emery, 1887) and *L. mutabilis* (Smith, 1861). *Zookeys* 59:47-60

Maruyama M, von Beeren C, and Witte V (2010b). Aleocharine rove beetles (Coleoptera, Staphylinidae) associated with *Leptogenys* Roger, 1861 (Hymenoptera, Formicidae) II.Two new genera and two new species associated with *L. borneensis* Wheeler, 1919. *Zookeys* 59:61-72

Moore J (2002). **Parasites and the behavior of animals.** Oxford University Press, New York, USA

Nash DR, Als TD, Maile R, Jones GR, and Boomsma JJ (2008). A mosaic of chemical coevolution in a large blue butterfly. *Science* 319:88-90

Nunn CL, and Altizier S (2006). **Infectious disease in primates.** Oxford University Press, Oxford, UK

Pamminger T, Scharf I, Pennings P, and Foitzik S (2011). Increased host aggression as an induced defense against slave-making ants. *Behavioral Ecology* doi: 10.1093/ beheco/ arq191

Poulin R, and Morand S (2000). The diversity of parasites. *The Quarterly Review of Biology* 75:277-293

Rettenmeyer CW, Rettenmeyer ME, Joseph J, and Berghoff SM (2011). The largest animal association centered on one species: the army ant *Eciton burchellii* and its more than 300 associates. *Insectes Sociaux*, 58: 281-292

Rigaud T, Perrot-Minnot M-J, and Brown JF (2010). Parasite and host assemblages: embracing the reality will improve our knowledge of parasite transmission and virulence. *Proceedings of the Royal Society London Series B* 277:3693-3702

Rutrecht ST, and Brown MJF (2008). The life-history impact and implications of multiple parasites for bumble bee queens. *International Journal for Parasitology* 38:799-808

Ruxton GD (2009). Non-visual crypsis: a review of the empirical evidence for camouflage to senses other than vision. *Philosophical Transactions of the Royal Society B-Biological Sciences* 364:549-557

Ruxton G, Sheratt T, and Speed M (2004). **Avoiding Attack: The Evolutionary Ecology of Crypsis, Warning Signals and Mimicry**. Oxford University Press, New York, USA

Scharf I, Pamminger T, and Foitzik S (2011). Differential response of ant colonies to intruders: attack strategies correlates with potential threat. *Ethology* 117:1-9

Schmid-Hempel P (1998). **Parasites in social insects**. Princeton University Press, Princeton, New Jersey, USA

Schönrogge K, Wardlaw JC, Peters AJ, Everett S, Thomas JA, and Elmes GW (2004). Changes in chemical signature and host specificity from larval retrieval to full social integration in the myrmecophilous butterfly *Maculinea rebeli*. Journal of Chemical Eclogy 30:91-107

Seevers CH (1965). The systematics, evolution and zoogeography of staphylinid beetles assoclated with army ants (Coleoptera, Staphylinidae). *Fieldiana Zoology* 47:139-351

Sheldon BC, and Verhulst S (1996). Ecological immunology: Costly parasite defences and trade-offs in evolutionary ecology. *Trends in Ecology & Evolution* 11:317-321

Steghaus-Kovac S (1994). **Wanderjäger im Regenwald- Lebensstrategien im Vergleich: Ökologie und Verhalten südostasiatischer Ameisenarten der Gattung *Leptogenys* (Hymenoptera: Formicidae: Ponerinae)**. Fachbereich Biologie Dissertation

Steiger S, Schmitt T, and Schaefer HM (2011). The origin and dynamic evolution of chemical information transfer. *Proceedings of the Royal Society Series B* 278:970-979

Stöffler M, Tolasch T, and Steidle JLM (2011). Three beetles—three concepts. Different defensive strategies of congeneric myrmecophilous beetles. *Behavior Ecology and Sociobiology* 65:1605-1613

Stowe MK (1988). Chemical mimicry. In: Spencer KC (ed) **The chemical mediation of coevolution**. Academic Press, New York, USA, pp 513-580

Symonds MRE, and Elgar MA (2008). The evolution of pheromone diversity. *Trends in Ecology and Evolution* 23:220-228

Thomas JA, Schönrogge K, Elmes GW (2005). Specializations and host associations of social parasites of ants. In: Fellowes MDE, Holloway GJ, and Rolff J (eds). **Insect Evolutionary Ecology: Proceedings of the Royal Entomological Society's 22nd Symposium**. CABI, Wallingford, UK

Thompson JN (2005). **The geographic mosaic of coevolution**. The University of Chicago Press, Chicago, USA

Thrall PH, Hochberg ME, Burdon JJ, and Bever JD (2007). Coevolution of symbiotic mutualists and parasites in a community context. *Trends in Ecology and Evolution* 22:120-6

van Zweden JS, and d'Ettorre P (2010). Nestmate recognition in social insects and the role of hydrocarbons. In: Blomquist GJ, Bagnères AG (eds) **Insect hydrocarbons: Biology, Biochemistry and Chemical Ecology**. Cambridge University Press, New York, USA, pp 222-243

Ward PS (2006). Ants. *Current Biology* 16:152-155

References

Wasmann E (1895) Die Ameisen-und Termitengäste von Brasilien. I. Teil. Mit einem Anhange von Dr. August Forel. *Verhandlungen der kaiserlich-königlichen zoologisch-botanischen Gesellschaft in Wien* 45:137–179

Wilson EO (1971). **The insect societies.** Harvard University Press, Cambridge, USA

Wilson EO (1990). **Success and dominance in ecosystems: the case of the social insects.** Ecology Institut Oldendorf/Luhe, GER

Witte V, Foitzik S, Hashim R, Maschwitz U, and Schulz S (2009). Fine tuning of social integration by two myrmecophiles of the ponerine army ant, *Leptogenys distinguenda*. *Journal of Chemical Ecology* 35:355-367

Witte V, Janssen R, Eppenstein A, and Maschwitz U (2002). *Allopeas myrmekophilos* (Gastropoda, Pulmonata), the first myrmecophilous mollusc living in colonies of the ponerine army ant *Leptogenys distinguenda* (Formicidae, Ponerinae). *Insectes Sociaux* 49:301-305

Witte V, Leingärtner A, Sabaß L, Hashim R, and Foitzik S (2008). Symbiont microcosm in an ant society and the diversity of interspecific interactions. *Animal Behaviour* 76:1477-1486

Wolinska J, and King KC (2009). Environment can alter selection in host–parasite interactions. *Trends in Parasitology* 25:236-244

Acknowledgements

Special thanks go to my supervisor Prof. Witte, who guided me markedly well throughout the dissertation. He took the time to assist me whenever I needed it in the field in Malaysia, in the laboratories or during writing at the LMU Munich. I am also very grateful to Prof. Foitzik, the former leader of the behavioural ecology working group. Special thanks also go to Sebastian Pohl for detailed and constructive comments on my manuscript drafts. I am grateful to the following persons who assisted me during field work or commented on earlier drafts of my manuscripts: Angelika Pohl, Deborah Schweinfest, Magdalena Mair, Tobias Pamminger, Ilka Kureck, Andreas Modlmeier, Stefan Huber, Max Kölbl, Hannah Kriesell, Carla Hegerl, Daniel Schließmann, Ricardo Caliari Oliveira, Julia Wickjürgen, and Nicola Flossdorf. Thanks go also to the following native English speakers who helped to improve my manuscripts: Tomer Czaczkes, William Robert Morrison III., Amanda Glaser, and Jared Lockwood. I am grateful for financial support from the DFG (Deutsche Forschungsgemeinschaft, project WI 2646/3). Thanks go also to my parents for their support. Last but not least, I want to thank my wife Sofia Lizon à l'Allemand who supported me throughout my doctoral research.

Behavioral ecology group 2011 at the LMU led by Volker Witte clothed in traditional Malayan Sarongs. Thanks to Ahmed who explained how to wear it. © Angelika Pohl

i want morebooks!

Buy your books fast and straightforward online - at one of world's fastest growing online book stores! Environmentally sound due to Print-on-Demand technologies.

Buy your books online at
www.get-morebooks.com

Kaufen Sie Ihre Bücher schnell und unkompliziert online – auf einer der am schnellsten wachsenden Buchhandelsplattformen weltweit! Dank Print-On-Demand umwelt- und ressourcenschonend produziert.

Bücher schneller online kaufen
www.morebooks.de

VDM Verlagsservicegesellschaft mbH
Heinrich-Böcking-Str. 6-8   Telefon: +49 681 3720 174   info@vdm-vsg.de
D - 66121 Saarbrücken   Telefax: +49 681 3720 1749 www.vdm-vsg.de

Printed by Books on Demand GmbH, Norderstedt / Germany